Next Generation HALT and HASS

Wiley Series in Quality and Reliability Engineering

Dr. Andre Kleyner
Series Editor

The Wiley series in Quality & Reliability Engineering aims to provide a solid educational foundation for both practitioners and researchers in Q&R field and to expand the reader's knowledge base to include the latest developments in this field. The series will provide a lasting and positive contribution to the teaching and practice of engineering.

The series coverage will contain but is not exclusive to:

- Statistical methods
- Physics of failure
- Reliability modeling
- Functional safety
- Six sigma methods
- Lead free electronics
- Warranty analysis/management
- Risk and safety analysis

A complete list of titles in this series appears at the end of the volume.

Next Generation HALT and HASS

Robust Design of Electronics and Systems

Kirk A. Gray
Accelerated Reliability Solutions, LLC, Colorado, USA

John J. Paschkewitz
Product Assurance Engineering, LLC, Missouri, USA

Registered Office
John Wiley & Sons, Ltd, The Atrium, Southern Gate, Chichester, West Sussex, PO19 8SQ, United Kingdom

For details of our global editorial offices, for customer services and for information about how to apply for permission to reuse the copyright material in this book please see our website at www.wiley.com.

Library of Congress Cataloging-in-Publication Data

Names: Gray, Kirk, author. | Paschkewitz, John James, author.
Title: Next generation HALT and HASS : robust design of electronics and systems/
    by Kirk Gray, John James Paschkewitz.
Description: Chichester, UK ; Hoboken, NJ : John Wiley & Sons, 2016. |
    Includes bibliographical references and index.
Identifiers: LCCN 2015044935 | ISBN 9781118700235 (cloth)
Subjects: LCSH: Accelerated life testing. | Electronic systems–Design and construction. |
    Electronic systems–Testing.
Classification: LCC TA169.3 .G73 2016 | DDC 621.381028/7–dc23
LC record available at http://lccn.loc.gov/2015044935

A catalogue record for this book is available from the British Library.

Set in 10.5/13pt Palatino by SPi Global, Pondicherry, India

1   2016

Contents

Series Editor's Foreword

The Wiley Series in Quality & Reliability Engineering aims to provide a solid educational foundation for researchers and practitioners in the field of quality and reliability engineering and to expand their knowledge base by including the latest developments in these disciplines.

The importance of quality and reliability to a system can hardly be disputed. Product failures in the field inevitably lead to losses in the form of repair costs, warranty claims, customer dissatisfaction, product recalls, loss of sales and, in extreme cases, loss of life.

Engineering systems are becoming increasingly complex, with added functions and capabilities; however, the reliability requirements remain the same or are even growing more stringent. Also the rapid development of functional safety standards increases pressure to achieve ever higher reliability as it applies to system safety. These challenges are being met with design and manufacturing improvements and, to no lesser extent, by advancements in testing and validation methods.

Since its introduction in the early 1980s the concept and practice of highly accelerated life testing has undergone significant evolution. This book *Next Generation HALT and HASS* written by Kirk Gray and John Paschkewitz, both of whom I have the privilege to know personally, takes the concept of rapid product development to a new level. Both authors have lifelong experience in product testing, validation and applications of HALT to product development processes. HALT and HASS have quickly become mainstream product development tools, and this book is the next step in cementing their place as an integral part of the design process; it offers an excellent mix of theory, practice, useful applications and common sense engineering, making it a perfect addition to the Wiley series in Quality and Reliability Engineering.

The purpose of this Wiley book series is not only to capture the latest trends and advancements in quality and reliability engineering but also to influence future developments in these disciplines. As quality and reliability science evolves, it reflects the trends and transformations of the technologies it supports. A device utilizing a new technology, whether it be a solar panel, a stealth aircraft or a state-of-the-art medical device, needs to function properly and without failures throughout its mission life. New technologies bring about new failure mechanisms, new failure sites and new failure modes, and HALT has proven to be an excellent tool in discovering those weaknesses, especially where new technologies are concerned. It also promotes the advanced study of the physics of failure, which improves our ability to address those technological and engineering challenges.

In addition to the transformations associated with changes in technology the field of quality and reliability engineering has been going through its own evolution by developing new techniques and methodologies aimed at process improvement and reduction of the number of design and manufacturing related failures. And again, HALT and HASS form an integral part of that transformation.

Among the current reliability engineering trends, life cycle engineering concepts have also been steadily gaining momentum by finding wider applications to life cycle risk reduction and minimization of the combined cost of design, manufacturing, warranty and service. Life cycle engineering promotes a holistic approach to the product design in general and quality and reliability in particular.

Despite its obvious importance, quality and reliability education is paradoxically lacking in today's engineering curriculum. Very few engineering schools offer degree programs, or even a sufficient variety of courses, in quality or reliability methods; and the topic of HALT and HASS receives almost no coverage in today's engineering student curriculum. Therefore, the majority of the quality and reliability practitioners receive their professional training from colleagues, professional seminars, publications and technical books. The lack of opportunities for formal education in this field emphasizes too well the importance of technical publications for professional development.

We are confident that this book, as well as this entire book series, will continue Wiley's tradition of excellence in technical publishing and provide a lasting and positive contribution to the teaching, and practice of reliability and quality engineering.

Dr. Andre Kleyner,
Editor of the Wiley Series in
Quality & Reliability Engineering

Preface

This book is written for practicing engineers and managers working in new product development, product testing or sustaining engineering to improve existing products. It can also be used as a textbook in courses in reliability engineering or product testing. It is focused on incorporating empirical limit determination with accelerated stress testing into a physics of failure approach for new product and process development. It overcomes the limitations, weaknesses and assumptions prevalent in prediction based reliability methods that have prevailed in many industries for decades.

We are especially grateful to the late Dr Gregg Hobbs for being the creator of HALT and HASS and a teacher and mentor.

We especially appreciate Dr Michael Pecht, the founder of CALCE at the University of Maryland, for his encouragement for writing this book and sharing CALCE material.

We would like to indicate our gratitude to our colleagues who provided support, input, review and feedback that helped us create this book. We thank Andrew Roland for permission to use his article *MTBF – What Is It Good For?* We would also like to thank Charlie Felkins for the pictures and drawings he provided and Andrew Riddle of Allied Telesis Labs for use of their case history. We are also grateful for the assistance of Fred Schenkelberg in providing support, contributions and promotion of this book.

We would like to thank Mark Morelli for material used in the book, as well as working with him early on implementing HALT and HASS at Otis Elevator, and Michael Beck for his support on implementing HALT and HASS, and access to information on DRBFM. We are grateful to Bill Haughey for introducing us to GD³ DRBFM and DRBTR, as well as to James McLeish for his support and work on Robust Design, Failure Analysis and GD³ source information.

We want to acknowledge Watlow and in particular Chris Lanham for providing opportunity to expand and apply our reliability knowledge, as well as Mark Wagner for his case history contribution to the Appendix.

Reliasoft granted us permission to use material in this book and we appreciate the support and encouragement from Lisa Hacker. We thank Linda Ofshe for her technical editing of early chapters, Richard Savage for his support and encouragement and Monica Nogueira at SAE International for her review of manuscript sections and resolving questions on copyrighted material.

Ella Mitchell, Liz Wingett and Pascal Raj Francois, who are our contacts at John Wiley & Sons, have guided us through the process of writing a technical book and all the details of manuscript development and preparation for publication.

List of Acronyms

ALT	Accelerated Life Testing
AMSAA	Army Material Systems Analysis Activity
AST	Accelerated Stress Tests
CALT	Calibrated Accelerated Life Test
CDF	Cumulative Distribution Function
CHC	Channel Hot Carrier
CND	Can Not Duplicate
CRE	Certified Reliability Engineer
DoD	Department of Defense
DFX	Design for X (Test, Cost, Manufacture & Assembly, etc.)
DFR	Design for Reliability
DFSS	Design for Six Sigma
DOE	Design of Experiments
DRBFM	Design Review Based on Failure Modes
DRBTR	Design Review Based on Test Results
DVT	Design Verification Test
ED	Electrodynamic (Shaker)
EM	Electromigration
ESS	Environmental Stress Screening
FEA	Finite Element Analysis
FIT	Failure in Time
FLT	Fundamental Limit of Technology
FMEA	Failure Modes and Effects Analysis
FMECA	Failure Modes, Effects & Criticality Analysis

FRACAS	Failure Reporting, Analysis, & Corrective Action System
GD³	Good Design, Good Discussion, Good Dissection
HALT	Highly Accelerated Life Test
HASS	Highly Accelerated Stress Screening
HASA	Highly Accelerated Stress Audit (Sampling)
HTOL	High Temperature Operating Life
HCI	Hot Carrier Injection
ICs	Integrated Circuits
LCD	Liquid Crystal Display
LCEP	Life Cycle Environmental Profile
MSM	Matrix Stressing Method
MTBF	Mean Time between Failures
MTTF	Mean Time To Failure
MWD	Measurement While Drilling
NBTI	Negative Bias Temperature Instability
NDI	Non Developmental Item
NFF	No Fault Found
NPF	No Problem Found
OEM	Original Equipment Manufacturer
ORT	Ongoing Reliability Test
PoF	Physics of Failure
PRAT	Production Reliability Acceptance Test
PTH	Plated Through Holes
PWBA	Printed Wiring Board Assembly
QFD	Quality Function Deployment
RoHS	Restriction of Hazardous Substances
RMA	Returned Material Authorization
RMS	Reliability, Maintainability, Supportability
RDT	Reliability Demonstration Test
SINCGARS	Single Channel Ground Air Radio Set
SPC	Statistical Process Control
TDDB	Time Dependent Dielectric Breakdown
VOC	Voice of the Customer
WCA	Worst Case Analysis

Introduction

This book presents a new paradigm for reliability practitioners. It is focused on incorporating empirical limit determination with accelerated stress testing into a physics of failure approach for new product and process development. This extends the basics of highly accelerated life test (HALT) and highly accelerated stress screens (HASS) presented in earlier books and contrasts this new approach with the limitations, weaknesses, and assumptions in prediction based reliability methods that have prevailed in many industries for decades. It addresses the lack of understanding of why most systems fail, which has led to reliance on reliability predictions.

Chapters 1, 2 and 3 examine the basis and limitations of statistical reliability prediction methods and shows why they fail to provide useful estimates of reliability in new products even if they are derivatives of previous products. It also addresses the prevailing focus on estimating life or reliability with metrics such as MTBF (mean time before failures) and MTTF (mean time to failure) and the misleading aspects of using these metrics in reliability programs. This includes difficulties and limitations in using field return data on previous products or results of reliability demonstration tests to derive an MTBF or MTTF estimate on new products. The section concludes with an assessment of practices in many reliability programs and shows how they can be inadequate, resulting in warranty claims, customer dissatisfaction and increased

Next Generation HALT and HASS: Robust Design of Electronics and Systems, First Edition.
Kirk A. Gray and John J. Paschkewitz.

cost to correct field problems. These typical practices include reactive reliability efforts conducted too late in product development to influence the design, success based testing that fails to find product weaknesses, and a focus on deliverable data to meet the customer's qualification requirements.

Chapter 4 proposes a new approach to ensuring product reliability. This begins with a focused risk assessment to anticipate potential failure modes and weaknesses based on changes from the current product knowledge base as well as new components and materials needed to meet customer needs. This assessment draws on knowledge of subject matter experts and tools to identify likely failure mechanisms and causes. These risks are then addressed with robust design to ensure sufficient margin to withstand the variability of anticipated operating environments and production strength variability. The robust design also considers prognostics and health management to detect degradation and wear out by monitoring key parameters during operation. This design approach is followed by phased robustness testing of prototypes using accelerated stress tests, including HALT, to find product limits and design margins as well as to identify design weaknesses. After the weaknesses have been identified, design changes to overcome the issues are completed and verified in HALT or accelerated stress tests.

With the empirical limits determined and weaknesses corrected, quantitative accelerated life test can be used to estimate reliability of selected components or assemblies where the operating environment stresses can be determined and applied. ALT provides indication of expected reliability in the reduced time available with today's shorter product development schedules. On systems with higher levels of integration, correctly identifying the combined stresses and accelerating them in a test becomes very difficult. So, validation testing at system level in the actual application may be needed to assess reliability and evaluate interfaces, which are often the source of reliability issues. Finally, production variability, process issues and supplier component variability need to be addressed with production screening tests and corrective action of issues discovered.

Chapters 5 and 6 detail the Highly Accelerated Life Test (HALT) from concept through process and planning to description of how to apply HALT. It also covers how to conduct failure analysis and ensure corrective action for the product weaknesses that are discovered. This includes selected stresses to apply in HALT, product configuration for test and applying thermal, vibration and power variation stresses, monitoring product operation and detecting failures and failure analysis after HALT.

Chapter 7 covers the use of production screening for electronics using Highly Accelerated Stress Screening (HASS) to find infant mortality issues and ensure the consistency and control of production processes. The HASS process is covered in detail, including precipitation and detection screens, stresses applied in HASS, the safety of screen process and verification of the HASS process. The effectiveness of HASS is discussed and transition to Highly Accelerated Stress Audit (HASA) sampling and cost avoidance are then covered.

Chapter 8 includes HALT and HASS examples to illustrate the application and effectiveness of discovering empirical limits, correcting design weaknesses and ensuring repeatable production processes. The section concludes with the benefits of HALT for software and firmware performance and reliability.

Chapter 9 covers the application of quantitative Accelerated Life Test (ALT) at component and subassembly levels when stresses can be correlated to the application environment and accelerated to levels between the operational level and the empirical limit of the product under test for the selected stresses used in the test. At higher levels of assembly, the combined stresses encountered in application become more difficult to apply and control to appropriate levels in an accelerated test. For these assemblies, validation testing in the application system at the prototype stage becomes necessary to evaluate interfaces and find potential problems that could not be discovered at the component or subassembly level.

Chapter 10 examines failure analysis, managing correction action and capturing learning in the knowledge base for access by follow-on project teams, allowing them to build on previous work rather than relearn it. This includes Design Review Based on Test Results (DRBTR) as a method for reviewing test results, deciding on corrective actions and tracking progress to completion and closure. Follow-up with production screening, ongoing reliability test during production and analysis of field data conclude the section.

Chapter 11 covers additional applications of the HALT methodology. These topics include:

- future of reliability engineering and the HALT methodology
- winning the hearts and minds of the HALT skeptics
- analysis of field failures in HALT
- test of no defect found units in HALT
- HALT for reliable supplier selection
- comparisons of stress limits for reliability assessments
- multiple stress limit boundary maps and robustness indicator figures
- focusing on deterministic weakness discovery will lead to new tools
- application of empirical limit test, AST and HALT concepts to products other than electronics

These areas help the reliability practitioner apply the HALT methodology and tools to solve problems they often face in both product development and sustaining engineering of current products.

The appendix includes data from case studies that illustrate the effectiveness of the HALT methods in improving product reliability.

1

Basis and Limitations of Typical Current Reliability Methods and Metrics

Reliability cannot be achieved by adhering to detailed specifications. Reliability cannot be achieved by formula or by analysis. Some of these may help to some extent, but there is only one road to reliability. Build it, test it and fix the things that go wrong. Repeat the process until the desired reliability is achieved. *It is a feedback process and there is no other way.*

David Packard, 1972

In the field of electronics reliability, it is still very much a Dilbert world as we see in the comic from Scott Adams, Figure 1.1. Reliability Engineers are still making reliability predictions based on dubious assumptions about the future and management not really caring if they are valid. Management just needs a 'number' for reliability, regardless of the fact it may have no basis in reality.

Next Generation HALT and HASS: Robust Design of Electronics and Systems, First Edition.
Kirk A. Gray and John J. Paschkewitz.
© 2016 John Wiley & Sons, Ltd. Published 2016 by John Wiley & Sons, Ltd.

Figure 1.1   Dilbert, management and reliability. Source: DILBERT © 2010 Scott Adams. Reproduced with permission of UNIVERSAL UCLICK

The classical definition of reliability is the probability that a component, subassembly, instrument, or system will perform its specified function for a specified period of time under specified environmental and use conditions. In the history of electronics reliability engineering, a central activity and deliverable from reliability engineers has been to make reliability predictions that provide a quantification of the lifetime of an electronics system.

Even though the assumptions of causes of unreliability used to make reliability predictions have not been shown to be based on data from common causes of field failures, and there has been no data showing a correlation to field failure rates, it still continues for many electronics systems companies due to the sheer momentum of decades of belief. Many traditional reliability engineers argue that even though they do not provide an accurate prediction of life, they can be used for comparisons of alternative designs. Unfortunately, prediction models that are not based on valid causes of field failures, or valid models, cannot provide valid comparisons of reliability predictions.

Of course there is a value if predictions, valid or invalid, are required to retain one's employment as a reliability engineer, but the benefit for continued employment pales in comparison to the potential misleading assumptions that may result in forcing invalid design changes that may result in higher field failures and warranty costs.

For most electronics systems the specific environments and use conditions are widely distributed. It is very difficult if not impossible to know specific values and distributions of the environmental conditions and use conditions that future electronics systems will be subjected to. Compounding the challenge of not knowing the distribution of stresses in the end - use environments is that the numbers of potential physical interactions and the strength or weaknesses of potential failure mechanisms in systems of hundreds or thousands of components is phenomenologically complex.

Tracing back to the first electronics prediction guide, we find the RCA release of TR-ll00 titled *Reliability Stress Analysis for Electronic Equipment*, in 1956, which presented models for computing rates of component failures. It was the first of the electronics prediction 'cookbooks' that became formalized with the publishing of reliability handbook MIL-HDBK-217A and continued to 1991, with the last version MIL-HDBK-217F released in December of that year. It was formally removed as a government reference document in 1995.

1.1 The Life Cycle Bathtub Curve

A classic diagram used to show the life cycle of electronics devices is the life cycle bathtub curve. The bathtub curve is a graph of time versus the number of units failing.

Just as medical science has done much to extend our lives in the past century, electronic components and assemblies have also had a significant increase in expected life since the beginning of electronics when vacuum tube technologies were used. Vacuum tubes had inherent wear-out failure modes, such as filaments burning out and vacuum seal leakage, that were a significant limiting factor in the life of an electronics system.

The life cycle bathtub curve, which is modeled after human life cycle death rates and is shown in Figure 1.2., is actually a combination of two curves. The first curve is the initial declining failure rate, traditionally referred to as the period of 'infant mortality', and the second curve is the increasing failure rates from wear-out failures. The intersection of the two curves is a more or less flat area of the curve, which may appear to be a constant failure rate region. It is actually very rare that electronics components fail at a constant rate, and so the 'flat' portion

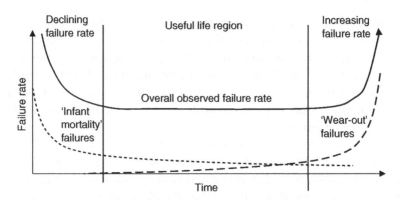

Figure 1.2 The life cycle bathtub curve

of the curve is not really flat but instead a low rate of failure with some peaks and valleys due to variations in use and manufacturing quality.

The electronics life cycle bathtub curve was derived from human the life cycle curves and may have been more relevant back in the day of vacuum tube electronics systems. In human life cycles we have a high rate of death due to the risks of birth and the fragility of life during human infancy. As we age, the rates of death decline to a steady state level until we age and our bodies start to fail. Human infant mortality is defined as the number of deaths in the first year of life. Infant mortality in electronics has been the term used for the failures that occur after shipping or in the first months or first year of use.

The term 'infant mortality' applied to the life of electronics is a misnomer. The vast majority of human infant mortality occurs in poorer third world countries, and the main cause is dehydration from diarrhea, which is a preventable disease. There are many other factors that contribute to the rate of infant deaths, such as limit access to health services, education of the mother and access to clean drinking water. The lack of healthcare facilities or skilled health workers is also a contributing factor.

An electronic component or system is not weaker when fabricated; instead, if manufactured correctly, components have the highest inherent life and strength when manufactured, then they decline in strength, or total fatigue life during use.

The term 'infant mortality', which is used to describe failures of electronics or systems that occurs in the early part of the use life cycle, seems to imply that the failure of some devices and systems is intrinsic to the manufacturing process and should be expected. Many traditional reliability engineers dismiss these early life failures, or 'infant mortality' failures as due to 'quality control' and therefore do not see them as the responsibility of the reliability engineering department. Manufacturing quality variations are likely to be the largest cause of early life failures, especially far designs with narrow environmental stress capabilities that could be found in HALT. But it makes little difference to the customer or end-user, they lose use of the product, and the company whose name is on it is ultimately to blame.

So why use the dismissive term infant mortality to describe failures from latent defects in electronics as if they were intrinsic to manufacturing? The time period that is used to define the region of infant mortality in electronics is arbitrary. It could be the first 30 days or the first 18 months or longer. Since the vast majority of latent (hidden) defects are from unintentional process excursions or misapplications, and since they are not controlled, they are likely to have a wide distribution of times to failure. Many times the same failure mechanism in which the weakest distributions may occur within 30 to 90 days will continue for the stronger latent defects to contribute to the failure rate throughout the entire period of use before technological obsolescence.

1.1.1 Real Electronics Life Cycle Curves

Of course the life cycle bathtub curves are represented as idealistic and simplistic smooth curves. In reality, monitoring the field reliability would result in a dynamically changing curve with many variations in the failure rates for each type of electronics system over time as shown in Figure 1.3. As failing units are removed from the population, the remaining field population failure rate decreases and may appear to reach a low steady state or appear as a constant or steady state failure rate in a large population.

In the real tracking of failure rates, the peaks and valleys of the curve extend to the wear-out portion of the life cycle curve. For most electronics, the wear-out portion of the curve extends well beyond technological obsolescence and will be never actually significantly contribute to unreliability of the product.

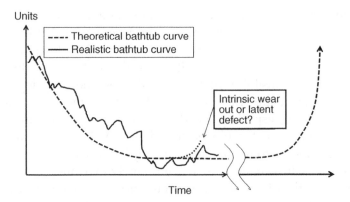

Figure 1.3 Realistic field life cycle bathtub curve

Without detailed root cause analysis of failures that make up the peaks of the middle portion of the bathtub curve, or what is termed the useful life period, any increase in failure rates can be mistaken as the intrinsic wear-out phase of a system's life cycle. It may be discovered in failure analysis that what at first appears to be an wear out mode in a component, is actually due to it being overstressed from a misapplication in circuit or unknown high voltage transients.

The traditional approach to electronics reliability engineering has been to focus on probabilistic wear-out mode of electronics. Failures that are due to the wear-out mode are represented by the exponentially increasing failure rate or back end of the bathtub curve.

Mathematical models of intrinsic wear-out mechanisms in components and assemblies must assume that all the manufacturing processes – from IC die fabrication to packaging, mounting on a printed wiring board assembly (PWBA) and then final assembly in a system – are in control and are consistent through the production life cycle.

Mathematical models must also include specific values of environmental stress cycles that drive the inherent device degradation mechanisms for each device, which may include voltage and temperature cycles and shock and vibration, which can interact to modify rates of degradation. The sum of all the stresses that a whole product is expected to be subjected to during its use is the life cycle environmental profile (LCEP).

The cost of failures for a company introducing a new electronics product to market are much more significant at the front end of the bathtub curve, the 'infant mortality' period, rather than the 'useful life' or 'wear-out' period in the bathtub curve. This includes the tangible and quantifiable cost of service and warranty replacements, and less tangible but real costs in lost sales due to perceptions of poor reliability in a competitive market.

There is little data or supporting evidence that in general electronics systems intrinsic life can be modeled and predicted, and this is especially true for the early life failures. The misleading approach of using traditional reliability predictions for reliability development will be discussed further in Chapter 2.

1.2 HALT and HASS Approach

The frame of reference for the HALT and HASS approach, reliability testing is as simple as the old adage that 'a chain is only as strong as its weakest link'. A complex electronics system is only as strong as its weakest or least tolerant or capable component or subsystem. Just like pulling on a chain until the weakest link breaks, HALT methods apply a wide range of relevant stresses, both individually and in combinations, at increasing levels in order to expose the least capable element in the system. If the failure mechanism causes catastrophic damage to a component, when a destruct limit is reached in HALT, makes it easier to isolate a weak link, identifying the weak link is easier to isolate. Operational weakness causing soft failures can be more challenging to isolate.

HALT (highly accelerated life test) is a process that requires specific adaptation when it is applied to almost any system and assembly. Because HALT is a highly adaptive process, the information given in this book will be general guidelines on how to apply HALT. How HALT is adapted to each type of product or assembly is unique to each, and presents a learning process for each different type of electronic and electromechanical system. It is advised that a company that plans to adopt HALT as a new process or a new user of HALT will have a significantly faster adoption and success in implementation if they have the guidance of an experienced HALT consultant. As in any newly introduced adoption of test new methods and techniques, there are many engineers and managers that will have misunderstandings of the process and the goals of HALT and HASS (highly accelerated stress screening). An experienced HALT consultant will have the data and knowledge to keep the focus on the adaptive application and relevance of the HALT process and future benefits of creating a robust, but not "over-designed" system. The period between the HALT application for reliability development of a new product and the observation of the actual reliability performance in the field with the lower failure rates as a result of HALT may take many months or longer. An experienced HALT consultant can be the champion of HALT during the additional expense of HALT during product development and before the actual benefits increased reliability due to HALT are realized in the field, as reduced warranty and early life field failures.

The same principles of testing to operational or destruct limits used for HALT of electronics circuit boards can be applied to electromechanical and mechanical systems for purpose of again finding the weakest link in the system applied to electromechanical and some mechanical systems. The main difference is in what stress stimuli are used. HALT for systems other than electronics is discussed further in Chapter 11.

The goal of HALT is to develop the stress margin capability and system strength to the fundamental limits of the current technologies during product development. The fundamental limit of the technology (FLT) is the stress level that cannot be exceeded without using non-standard electronics materials or methods.

HALT is used to find stress limits and design weaknesses that could decrease field reliability, and is best performed during design and development phase. HASS is an ongoing application of combinations of stresses, defined from stress limits found empirically during HALT to detect any latent defects or reduction in the design's strength introduced during mass manufacturing.

Only after a system weakness is discovered can it be investigated and its significance and relevance to reliability be determined. Occasionally a weakness found in HALT is evaluated and not considered a risk of causing field failures. The opportunity to evaluate a weakness only comes when you find the stress limits. If the product is not tested to stress limits or failure, there is nothing to evaluate for potential reliability improvement.

HALT is becoming more widely adopted by electronics companies in the 21st century, although it is also more a current industry buzzword that may be used for marketing promotion than a process for actual improvement of electronics systems by increasing stress-strength margins. Suppliers of some subsystems in the IT hardware industry, such as power supplies, memory, or graphics display devices may use HALT, but the specifics of what is called a HALT can vary widely. It has been the author's experience that many purportedly using HALT may do stress tests, but only stress to a predetermined stress level that someone has arbitrarily determined is 'good enough'. One valuable result of HALT is the comparison of stress limits found between samples of the same product in HALT. Without finding empirical limits they will not be able to compare limits between samples of the same product. Wide distributions of strength seen as large differences in empirical operation or destruct limits can be an indication of inconsistent manufacturing at some level of the product.

One of the author's consulting clients had been performing HALT for many years on their products, yet when asked what the thermal operational limit was for one product of concern they admitted that they did not know because the HALT was stopped at 80°C because that was 'good enough'. Without finding a thermal operational limit, they missed discovering an important and revealing comparison of the operational limits between samples.

1.3 The Future of Electronics: Higher Density and Speed and Lower Power

Moore's Law, the projection that Gordon Moore made in 1965 that the number of components on an integrated circuit would approximately double every two years, has become an industry expectation for new component designs. The increase in densities of integration, reduction of feature sizes in integrated circuits and new packaging technologies introduces new fabrication and use physics that drive failure mechanisms and this is expected to continue for the foreseeable future.

Other changes in electronics materials may be implemented from concerns of the impact of electronics on the earth's environment. The change in going from using leaded solders to lead-free solders, and restricting the use of flame retardants are two examples of changes required by the directive on the restriction of the use of certain hazardous substances in electrical and electronic equipment. The directive was made by the European Union in 2002 for all electronics sold there and has been adopted worldwide. It is now commonly abbreviated as RoHS (Restriction of Hazardous Substances).

In the design and development of electronics, all of the changes and the rapidly increasing density and complexity of devices and systems make modeling each potential failure mechanism a moving target. Soon after a model of a new technology or failure phenomena is introduced, new materials and new technologies change the underlying physics of the causes of wear-out failures in devices or systems during their use.

The reliability of an electronics system is a phenomenologically complex issue. Prediction models do not include all the potential design and manufacturing errors or process deviations that may affect device and system reliability. Models of electronic component failure mechanisms that are used for reliability predictions are – and must be – based on the assumptions of the manufacturing processes being consistent and capable at all levels of system fabrication and throughout the manufacturing life cycle.

1.3.1 There is a Drain in the Bathtub Curve

The life entitlement of today's electronics components and systems with no moving parts far exceeds the life needed before a system is replaced by a newer more capable system. Technological obsolescence comes faster for today's electronics systems and they are replaced long before their life is consumed. The timescales between intrinsic wear-out modes of active devices and technological obsolescence of a system is significant in the vast majority of electronics. Because of the large difference in the timescales between obsolescence and wear-out of components and assemblies, wear-out mechanisms in electronics systems will never be observed. Also, because of the long life entitlement of electronics, using a small percentage of the fatigue life of electronics during HASS in production in order to find latent defects leaves more than enough life for the system to be shipped as new and to exceed the period of its technological obsolescence.

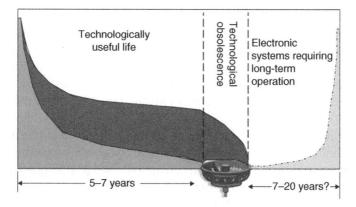

Figure 1.4 The 'drain' of technological obsolescence in the life cycle bathtub curve

There are electronic devices, such as batteries, that still have a relatively limited life compared to most circuit components, and for that reason are typically designed to be easily replaced. With few exceptions, intrinsic wear-out mechanisms on most components have not been shown to contribute to electronics systems unreliability during the first years of operation. Almost all failures of electronics in the first years are due to assignable causes, such as overlooked low design margin, an error in manufacturing, or misapplication or abuse by the customer, among many potential causes. An illustration of the electronics bathtub curve and the 'drain' of technological obsolescence is shown in Figure 1.4.

The vast majority of the costs of failures for almost all electronics manufacturers come in the first few years in the product's life. The left side of the bathtub curve shows a declining failure rate.

It is during the time of warranty coverage that company must pay for system repair or replacement costs for failed units. Failures during and shortly after the warranty period may also be a much greater a contributor to the financial loss for an electronics OEM as a result of lost sales from the market's perception of lower reliability. Loss of market share, and therefore unit sales, may be much greater than the material and service warranty costs, but since it is difficult to quantify lost sales, the actual monetary losses may never be known.

Technological obsolescence occurs at a much faster rate than intrinsic mechanisms in electronics and systems that lead to wear-out for most electronics systems and it is especially true in consumer electronics. It can be argued that most failures of electronics systems are due to errors in the design, manufacturing, or customer misuse or abuse.

Failure Prediction Methodologies (FPM) are more relevant for mechanical systems (i.e. motors, gears, switches) which can have a more limited life due intrinsic to friction and fatigue damage wear out mechanisms. In mechanical systems wear-out, the lifetime use can be modeled from physical measurements of material consumed, change in torque resistance or current draw, or other relevant measurements. The models can then be used to determine whether the wear-out duration of the mechanical device is adequate for the required life or mission requirement. If the intrinsic life is limited by the consumption of material (as in mechanical bearings) the reservoir of material can be increased meet the life requirements.

Technology has changed significantly in electronics in the past decades as IC densities and metallization line widths have continued to shrink, and lower voltage, faster ICs with more functionality are introduced every year. Yet, in the field of electronics reliability engineering, little has changed. The concepts and theories based on MIL HDBK-217 are still widely used, even though MIL HDBK-217 was removed as a government reference document and has not been updated or republished since the last revision ('F', notice 2) in 1991. Much of the data on failure rates of components, such as fans, is outdated by decades and has little relevance to today's electronics. Because of decades of reliance on handbook-based or 'cookbook' reliability predictions and invalid assumptions regarding temperature and component life, there is a continued perception that the higher the temperature at which electronics are operated, the faster the system will use up its 'life entitlement' and the sooner it will fail – regardless of well-documented evidence to the contrary.

1.4 Use of MTBF as a Reliability Metric

Traditional reliability engineering methods have focused on producing a quantitative reliability prediction based on time. The most widely used metrics in reliability are the terms 'mean time between failures' (MTBF) for repairable systems and 'mean time to failure' (MTTF) for non-repairable systems. MTBF is a single average of the total number of hours a set group of systems have operated between repairs or with MTTR until the first failure. Historically traditional reliability predictions use this single number to describe what can be very different distributions of failure rates. Because it is an average number, without more information it is not very useful for under-standing the probability of failures based on use or age of the prod-uct. It is a broad statistic that should not be used as a metric for defining reliability design goals or for field analysis of failures and warranty returns.

MTBF is a poor metric for providing information on the reliability of any system. It is derived from a very simple equation:

$$MTBF = \theta = \int_0^\infty t \times f(t)dt \qquad (1.1)$$

If we have 40 units that all run for 100 hours and right at the end of 100 hours one of the units fails, we can calculate the MTBF as follows. First determine the total hours that all the units operated. It is a very simple calculation, 40 units times 100 hours is 4000 hours. Next divide the total operating hours by the number of failures. One failure makes for a simple example: dividing by one the resulting MTBF is equal to 4000 hours.

The following section 1.5, written by Andrew Rowland who is a Certified Reliability Engineer (CRE), explains how the same MTBF number is calculated for three significantly different distributions and reliability risks.

1.5 MTBF: What is it Good For?

1.5.1 Introduction

The mean time between failure (MTBF) is arguably the most prolific metric in the field of reliability engineering. It is used as a metric throughout a product's life cycle, from requirements, to validation, to operational assessment. Unfortunately, MTBF alone doesn't tell us too much.

It's not that MTBF is a bad metric; it is just an incomplete metric, and as an incomplete metric it doesn't lend itself to risk-informed decision-making. The real problem is not with the MTBF, but with the implicit assumption that failure times are exponentially distributed.

1.5.2 Examples

To illustrate how relying on the MTBF can be misleading, let's look at two examples. In these examples we will assume that the failure times are Weibull distributed. The Weibull distribution is popular in reliability engineering and the exponential is a special case of the Weibull. From the literature, we know that the probability density function and survival (or reliability) function of the Weibull can be expressed as:

$$f(t) = \left(\frac{\beta}{\eta}\right)\left(\frac{t}{\eta}\right)^{\beta-1} e^{-\left(\frac{t}{\eta}\right)^{\beta}} \tag{1.2}$$

$$S(t) = e^{-\left(\frac{t}{\eta}\right)^{\beta}} \tag{1.3}$$

We also recall that the mean of a Weibull distributed variable can be estimated as:

$$MTBF = \eta\Gamma\left(1 + \frac{1}{\beta}\right) \tag{1.4}$$

In these functions, η is referred to as the scale parameter and β the shape parameter.

1.5.2.1 Example 1

Consider three items; item A, item B and item C. Perhaps the goal is to select one of these items for our design, and the requirement is to have a 90 hour MTBF or greater. All three items have an MTBF of 100 hours. So, from a reliability perspective, which is the item to choose?

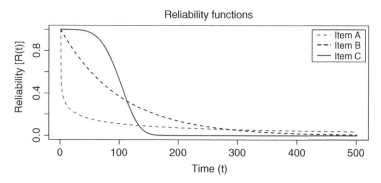

Figure 1.5   Reliability functions for item A, item B and item C

Table 1.1   Reliability at 100 Hours
for item A, item B and item C

Item	R(100)
Item A	0.109 (10.9%)
Item B	0.367 (36.7%)
Item C	0.521 (52.1%)

Under the implicit assumption that failure times are exponentially distributed, we might conclude that any of the three is acceptable, reliability-wise. All three satisfy the 90 hours MTBF requirement. However, let's look a little deeper into the 100 hour MTBF and see if we still agree that any of the three is acceptable.

Let's take a look at the reliability over time of each item. Figure 1.5 shows the reliability function over 500 hours for each of these items. Clearly, the reliability of these items is not the same. Given that each item has an MTBF of 100 hours, what is the reliability at 100 hours? Table 1.1 summarizes the 100 hour reliability for each item. Once again, we can see a large difference between the three items.

Another way to compare these three items is via the hazard, or failure, rate. Figure 1.6 shows the hazard function for each item. The 'bathtub' curve is a plot of hazard rate versus time. Thus, Figure 1.6 shows the 'bathtub' curve for each item. Clearly the hazard rate behaviors are very different for these items.

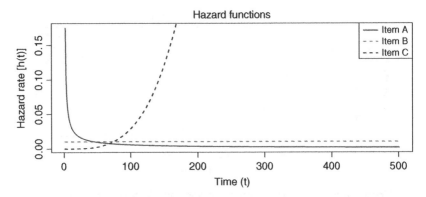

Figure 1.6 Hazard functions for item A, item B and item C

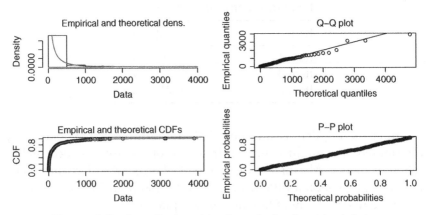

Figure 1.7 Item D: Graphical analysis of survival data

1.5.2.2 Example 2

Consider another situation where we have three items; item D, item E and item F. Presume for a moment that we have all of the data used to derive the MTBF statistic for each item. The first thing we might do is graphically explore the data. Figure 1.7 shows a set of plots commonly used in graphical analysis of survival data for item D. Let's look at the histogram in the upper left corner. We see that the distribution is heavy-tailed, indicating failure times are not exponentially distributed.

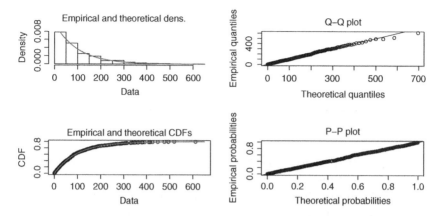

Figure 1.8 Item E: Graphical analysis of survival data

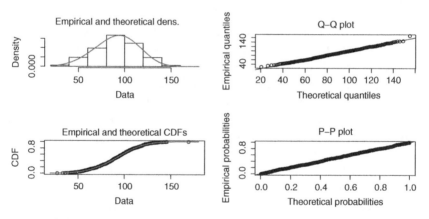

Figure 1.9 Item F: Graphical analysis of survival data

Compare the histogram in Figure 1.7 to that in Figure 1.8 for item E and Figure 1.9 for item F. Clearly the distribution of failure times differs among these three items. Yet all three items have the same MTBF. Perhaps we need to look a bit closer at the data! Now that we've graphically analyzed the data and concluded that we may be looking at different populations, we decide to fit the data to a distribution and estimate the parameters.

Table 1.2 Estimated parameters for item D, item E and item F

Item	Eta	Beta	MTBF
Item D	101.42	0.478	220.7
Item E	107.73	1.000	107.7
Item F	100.84	4.524	92.0

Our goal, then, is to estimate the value of β and η for each item. We use the fitdist function from the R [1] package, fitdistrplus [2] which uses maximum likelihood to estimate the parameters. The results for these three populations are summarized in Table 1.2. We can see from these results that the populations are not the same, although all three items satisfy our 90 hours MTBF requirement.

Now that we're confident that we're dealing with three different populations, all with the same MTBF, what is the implication of selecting one item over another? Since we fit the data to a Weibull distribution, we know the shape parameter (β) determines the region of the 'bathtub' curve. With a $\beta < 1$, we are in the early life region, a $\beta = 1$ puts us in the useful life region, and a $\beta > 1$ indicates wear-out. In other words, item D is dominated by early-life failure mechanisms, item E is dominated by useful life failure mechanisms, and item F by wear-out.

As we did with the first example, let's look at the reliability function for these three items.

Figure 1.10 shows the reliability functions. Similar to the first example, we see the reliability functions are not the same as we would expect from our assessment of Figures 1.4, 1.5 and 1.6.

Let's assume we are interested in the reliability at 50 hours. The reliability at 50 hours for the three items can be found in Table 1.3. We see a dramatic difference in the reliabilities and, interestingly, the item with the highest 50 hour reliability is the item with the lowest MTBF.

We can also look at plots of the hazard function for these three items. These hazard functions are plotted in Figure 1.11 over 500 hours. We see different hazard rate behaviors as we expected from our assessment of the β values we estimated earlier.

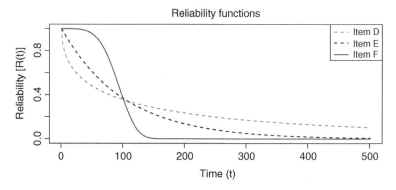

Figure 1.10 Reliability functions for item D, item E, and item F

Table 1.3 Reliability at 50 hours for item D, item E, and item F

Item	R(50)
Item D	0.490 (49.0%)
Item E	0.645 (64.5%)
Item F	0.959 (95.9%)

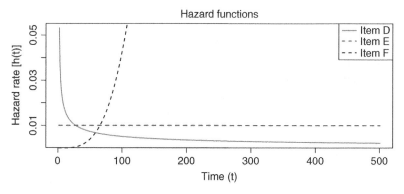

Figure 1.11 Hazard functions for item D, item E, and item F

1.5.3 Conclusion

Hopefully we've come to understand that stating an MTBF value with no other information doesn't really tell us much about the reliability of an item. Neither does it tell us if the item truly satisfies our reliability needs. We saw in one example three items with the same MTBF, but most definitely with different reliability behaviors.

In the second example, we looked at three items with different MTBF. Once again, we saw the reliability behaviors of these items were different. In this example, we saw the item with the largest MTBF having a 50 hour reliability almost half that of the item with the lowest MTBF.

Without an understanding of the reliability characteristics that is more complete than simply MTBF are we making good, risk-informed decisions? Selecting item A or item D, we can expect to see high rates of failure during validation, reliability growth testing or, worse yet, early in customer ownership. If we warrant our product, we can expect large warranty costs associated with items A or D. Given the competing requirements we need to satisfy, we may need to select item A or item D. If we only know the MTBF will we put the necessary barriers in place, such as screening, to minimize the risk?

At the other end of the 'bathtub' curve, if we select item C or item F, our validation or reliability growth testing may not test far enough into wear-out to surface failures. Will we develop a preventive maintenance program for these items to minimize the risk?

MTBF is ingrained in the reliability community as well as throughout most companies. It is unlikely that we will ever see the end of MTBF. Ultimately it comes down to us, as reliability engineers, to understand the limitations of MTBF and educate those around us to its shortcomings. If the reliability community gets in lock-step, we can be the tugboats that change the ship's heading.

~~~~~~~~

The use of MTBF will likely continue along with other misunderstandings of the realities of actual field unreliability since real reliability information that is needed to clarify the rates and causes of field unreliability of most electronics products will never be disclosed. The reason is that publishing the real causes of unreliability of electronics risks potentially very costly liability and litigation and market share loss for a electronics producer in a competitive marketplace. The change that is needed in electronics reliability will largely come from engineers who have observed and understand the root causes of field failures, not theoretical component failure rates or assumptions of wear-out mechanisms, to change the fundamental approach from developing reliable systems using theoretical assumptions to an approach using deterministic empirical discovery of weaknesses in an electronics and electromechanical system.

1.5.4 Alternatives to MTBF for Specifying Reliability

Fred Schenkelberg, an experienced reliability engineering and management consultant, is so passionate and determined to help remove the term MTBF from reliability engineering that he has created a website, 'No MTBF' that is dedicated to using better metrics than MTBF to define reliability requirements. Fred has written the following regarding the use of MTBF as a reliability metric.

> 'MTBF is often used to represent product life. It is neither complete nor sufficient. Product life or reliability has four elements: function, probability, duration and environment. MTBF is only the probability and assumes (in most cases) the duration does not matter, or worse is not even stated.
>
> As an alternative, use reliability directly. State the probability of success over a specified time frame, along with the functions (leads to understanding of product failure definition) and environment. The function and environment are often abbreviated, i.e. a respirator provides life support breathing in North American intensive care facilities. The details of the functions and environment are often well stated in product development and marketing documents.
>
> The probability and duration may include multiple statements. One statement might be for the critical period of the product life. For example, since products that experience failure during first use damage the product brand significantly, we may want to have a very high probability of success during the first 3 months of product use. Say, 99.99% reliability over first 3 months of use.

The warranty period may be duration of interest. In that case the statement for that period would be 98% reliability over the 1 year warranty period. And, the design life (how long the product should last and provide value to the customer) might be stated as 90% reliability over 5 years.

The early failures focus on component, assembly, shipping and installation sources of product failure. The warrant period and reliability is of interest as a business liability. The design life focuses on the longer term failure mechanisms.

Therefore, move away from a partial statement concerning product reliability. Make full use of clear statements of expectations (goals) and measures.' [3]

1.6 Reliability of Systems is Complex

The overall reliability of electronic assemblies and systems is a phenomenologically complex interaction of materials, manufacturing processes and end user applications and the broad potential variations in each of these factors.

If all the functions of design and assembly are performed correctly and if the system is used as intended, it will likely operate without failure until it is technologically obsolete. The pace of electronics technology is increasing and there is no reason to believe that it will slow down. The time for developing reliability in new electronics systems has become and will continue to be shorter. A faster method of ensuring the reliability of electronics systems is needed and will be required for meeting the market expectations and demand.

Gregg Hobbs, with his development of HALT and HASS, derived a much more efficient approach to reliability development using empirical limits under step stress testing to discover elements of a new design that could become a field reliability risk.

The most valuable time for the creation of a reliable new electronics system is during the design phase when the costs of changes are the lowest. A robust and reliable design provides a higher tolerance to extremes of environmental stress and potential abuse of the product, as well as creating margins that allow a higher tolerance to variations in the manufacturing processes.

At any point in the manufacturing process a latent defect can be introduced unknowingly and take a product that had been reliable to one that has poor reliability. There are some exceptions, but for most electronics components and systems the life entitlement – that is the length of time it functions before inherent wear-out mechanisms driven by fatigue or chemical reactions result in failure – is much longer than the time at which it is retired because it is technologically obsolete. Most electronics systems have a significant margin between the life entitlement of a properly designed and properly manufactured electronic system relative to that the product is technologically obsolete.

At each manufacturing level of an electronic system there can be variations in the quality and consistency of materials and processes used in the production of systems. Some common latent defects that cause electronics systems to fail can be introduced at each subsequent level of assembly, as shown in Figure 1.12.

For the vast majority of electronics systems, it can be very difficult, if not impossible, to know the life cycle environmental profile

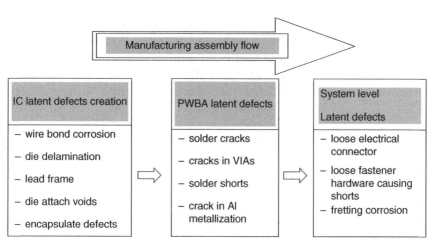

Figure 1.12 Examples of where latent defects are introduced during assembly fabrication

(LCEP) that any particular system will be exposed to during its use. Even if the LCEP is determined for a system, there may be a new use or application that was not considered during product development and that has significantly different environmental conditions. A good example would be a portable video projector. One population of a particular system may be attached to a room ceiling and have much less shock, vibration, and thermal cycling environmental stress. Another portion of the projectors purchased will be transported regularly by the user to various locations and will have many more mechanical shocks and vibration events, as well as thermal stress variations, compared to ceiling mounted projectors, yet the warranty and reliability expectations of the end user will be the same. The projector LCEP will have a wide distribution of conditions between environments yet the expectations for reliability and warranty coverage are the same regardless of the end use environmental conditions.

1.7 Reliability Testing

Reliability testing and assessment has been strongly influenced by FPM as shown in Figure 1.13.

Reliability predictions from FPM guides such as MIL-HDBK-217F, are based on the invalid assumption that the Arrhenius equation applies for many wear-out modes in semiconductors and other electronics components and has resulted in unnecessary costs in additional cooling and the belief that thermal derating during design

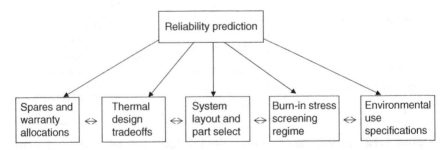

Figure 1.13 Impact of reliability tasks on electronics. Source: Adapted from Pecht and Nash, 1994

provides longer life. It also influences testing regimes with the belief that testing with steady state elevated temperature can provide a quantifiable acceleration of intrinsic wear-out mechanisms in electronics assemblies. There has been no data or evidence to support these beliefs.

Thermal and vibration stress has long been known to be a very useful stress to find latent defects in electronic hardware. In 1982, Hughes Aircraft published a guide entitled *Stress Screening of Electronic Hardware*, which was an early guide on using environmental stress screening (ESS). The objective of the guide was 'to develop methodologies and techniques for planning, monitoring and evaluating stress screening programs during electronic equipment development and production.'

One very interesting aspect of the development of the environmental stress screening curves shown in the Hughes Aircraft guide was the comparisons of the effectiveness of different stress stimuli used to precipitate the latent defects to patent or detectable defects.

In the Hughes ESS guide they confirm that thermal cycling stress screens and random vibration screens were generally the most effective screens for finding latent defects in electronics systems. They also acknowledge that the industry consensus was that the effectiveness of thermal cycling screens increases with wider temperature ranges and greater rates of change. Additionally it illustrated the industry knowledge that random or broadband vibration is more effective than single or sweep frequency sine vibration.

The vibration regime of a 6 Grms (gravity root mean squared) ESS profile presented in the government publication *Navy Manufacturing Screening Program* (see Figure 1.14) was intended to be a guideline. The 6 Grms vibration profile became the de facto standard auto spectral density (ASD) profile and was applied generically to all systems. Although ESS was a useful new method for finding latent defects, it may have been ineffective for some systems by using too low of stress levels to find defects, and for other systems it may have used stresses severe enough to shorten the products usable life.

HASS processes, like ESS processes, have the identical goal of finding latent defects. The most significant difference between HASS and ESS is how stress levels for a production stress screening process are determined.

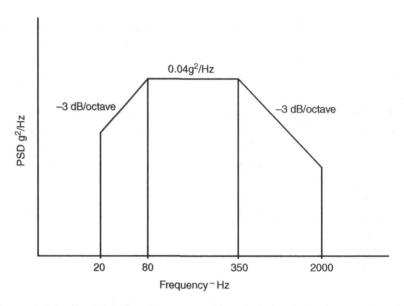

Figure 1.14　The ESS vibration power spectral density spectrum guideline from NAVMAT 9492 (US Navy)

The levels of stress for ESS were determined by 'stress screening strength' curves derived from industry consensus regarding the levels of stress needed to precipitate to detection a percentage of latent defects that would be expected per number of components in an assembly or subsystem being screened. In comparison the levels of stress used for HASS encompass a variety of stresses before product is shipped and is uniquely developed based on the product's empirical strength limits found in the HALT process.

In fact, the UUT (unit under test) in an ESS regime was not typically powered or monitored during the application of stress. Powering and functionally monitoring the UUT is another significant difference between ESS and HALT and HASS. In HALT and HASS, the product may be power cycled and briefly off during the stress application, but should be operating and its function monitored as much as possible during the process.

Many types of latent defects in electronics systems that are likely to become field failures may only be detectable during the application of stress. An example could be a ball grid array (BGA) solder joint that may have a 100 per cent fracture across the ball, but the surfaces without stress make contact, completing a conduction path that allows the product to operate normally. Only when the surfaces separate under thermomechanical stress or vibration is the conduction path open, which results in a detectable failure if the circuit operation is being monitored at the same time. If tested before and after a HALT without operational monitoring during the application of stress, many of the latent defects and weaknesses could go undetected. In some cases may be necessary to stress a product beyond operational levels for it to provide sufficient acceleration for a latent defect, followed by a lower stress level to operate the UUT for detection of the defects for an effective HASS.

HALT and HASS methods have provided documented cases of detecting operational reliability issues in the field. Many times a marginal system may have a degraded operational reliability from intermittent 'soft failures'. Soft failures are defined as the system failing but recovering normal operation when reset or power cycled. Soft failures may be more prevalent than catastrophic failures in the field, but unless they occur frequently, they may not be recorded, since no hardware needs to be replaced to return the unit to operation.

Many readers may have experienced a screen 'lock up' or 'blue screen of death' operational failure on a personal computer or other personal digital hardware. It can be an annoyance or worse, but it is usually a reason to return the device if it recovers and functions normally when we reboot or power cycle the system. If these 'soft' failures occur frequently enough, the user may return the unit to the manufacturer. It is often that due to the intermittent nature of the failure, the manufacturer will likely declare it 'no defect found' from the limited failure analysis it may have when returned. But the user's perception of overall poor reliability or quality will likely be told to others and may result in the purchase of a different brand when it comes time to upgrade.

As digital systems have been pushing up bus speeds to the gigahertz range and beyond, thermal stress, stepping up the clock frequency and voltage margining to limits will provide more sensitive discriminators to increase the probability of finding software and marginal signal integrity issues that result in operational reliability issues.

Variation in manufacturing causes variations in parametric performance from sample to sample, or lot to lot of electronic components and assemblies. The parametric variations at each assembly level stack up and can lead to timing and signal integrity failures. If the signal integrity is near the margin of failure at room temperature, it may become an intermittent soft failure if operated at higher or lower temperature, but still within the specifications of a design.

Soft failures due to marginal signal integrity can be some of the most challenging to find. It may take hundreds of operational cycles on many samples to reproduce the fault at nominal room conditions. For many engineers performing reliability testing, the potential benefit of stimulating variations of signal propagation and timing may never be realized because the fear that failures from HALT are due to stress levels that the system will never experience in its end use environment, and this is irrelevant and therefore will lead to "over-design".

If the fears of over-engineering a system are set aside long enough to perform a HALT on a new product, the HALT may demonstrate that a design is very robust and has significant margins. When a weakness is found in a properly run HALT, its relevance to field reliability can be determined and, in most cases, it is relevant. Finding the stress limits provides an opportunity to find and improve the weaknesses that may result in field unreliability, and to establish benchmarks for similar products. Testing to environmental specifications, or expected worse-case conditions, will not accelerate or provide a faster rate of cumulative fatigue over the fielded products that end up being used in a worst case environment. The point of accelerated testing is to find latent defects in electronics that result in failures, so that your customers do not find them. Worst case stress testing will find weaknesses and latent defects in the same time period for products being subject to worst-case end-use environments.

The only way to confirm if a weakness found in HALT is relevant to the field is to ship the units without improving the weakness and wait for failures. Of course this is a significant economic risk for most companies, and for most users of HALT the additional expense of improving weaknesses and possibly "over-designing" a product is much smaller that the potential costs of field failures if the weakness is not addressed.

1.8 Traditional Reliability Development

Since the early days of solid state electronics, reliability engineers have been taught that the dominant cause of hardware unreliability comes from component failures and that the reliability of components can be as much as doubled for each 10°C reduction in temperature. This belief was a fundamental tenet of the *U.S. Military Handbook 217* (MIL-HDBK-217),[1] the first document on the reliability prediction of electronic components [5]. While there is no empirical data to support this belief, the concept has persisted and has made its way into other reliability prediction handbooks, such as Telcordia SR-332 (formerly Bellcore), PRISM, FIDES and the Chinese GJB-299. These prediction methods rely on the analysis of insufficient failure data collected from the field, and they assume that the components of a system have inherently constant failure rates that can be derived from the collected data. These methods assume that such constant failure rates could be tailored by independent 'modifiers' to account for various quality, operating and temperature parameters.

In the 1990s, with a host of studies conducted by the National Institute of Standards and Technology (NIST) [6], Bell Northern Research [7], the U.S. Army [8], Boeing [9], Honeywell [4], Delco [10], Ford Motor Co. [11], and British Aerospace [12], it became clear that the approach propagated by these handbooks has been damaging to the industry and that a change was needed. The consensus is now that these methods and this type of approach should never be used, because they are inaccurate for predicting actual field failures and they provide highly misleading predictions, which can result in poor designs and poor logistics decisions [13]. Although most of these handbooks have been discontinued and are no longer used by the U.S. military, a few manufacturers of electronic components, printed wiring and circuit boards, and electronic equipment and systems even today still subscribe to the traditional reliability prediction techniques (e.g. MIL-HDBK-217 and its progeny) in some manner, although sometimes unknowingly.

---

[1] The last version of Mil-HDBK 217 was revision 'F', in 1995. Since then the document has been cancelled and not updated. Regardless of the fact that the predictions are inaccurate and misleading, it continues to be used have an influential role in reliability engineering.

Electronics systems, especially in the consumer products, have undergone a relatively rapid increase in technological features and benefits. For example, in less than 10 years, the cellular phone industry has gone from a simple portable unit that makes and receives calls to the current smart phones, which are small handheld computers.

When using models to estimate the life entitlement of a component or system certain assumptions must be made that the manufacturing processes are consistent with little variation in its fit or function. Properly manufactured components that are not in a marginal circuit are generally not the cause of the vast majority of hardware failures.

The 'life entitlement' of today's microelectronic components is not known and may never be known, but for most applications it is long beyond any required use time and almost always will reach far beyond the time when the component becomes obsolete.

Bibliography

[1] R Development Core Team, *R: A Language and Environment for Statistical Computing*. Vienna, Austria: R Foundation for Statistical Computing, 2009

[2] Delignette-Muller, M.L, Pouillot, R., Denis, J. and Dutang, C. *fitdistrplus: help to fit of a parametric distribution to censored or non-censored data*. 2013

[3] Schenkelberg, F. No MTBF-Actions. *NOMTBF.COM*. [Online] 2014. [Cited: 5 May 2014.] http://nomtbf.com/actions/.

[4] Pecht, M.G., Nash, F.R. *Predicting the Reliability of Electronic Equipment*, 1994, Proceedings of the IEEE, pp. 992–1004.

[5] US Department of Defense. *MIL-HDBK-217: Military Handbook for Reliability Prediction of Electronic Equipment, Version A*. 1965.

[6] Kopanksi, J., Blackburn, D.L., Harman, G.G., Berning, D.W. *Assessment of Reliability Concerns for Wide-Temperature Operation of Semiconductor Device Circuits*, Albuquerque, NM: Transactions of the First International High Temperature Electronics Conference, 1991.

[7] Witzmann, S. and Giroux, Y. *Mechanical Integrity of the IC Device Package: A Key Factor in Achieving Failure Free Product Performance*. Albuquerque, NM: Transactions of the First International High Temperature Electronics Conference, pp. 137–142, 1991.

[8] Cushing, M.J. US Army Reliability Standardization Improvement Policy and its Impact. *IEEE Transactions on Components, Packaging, and Manufacturing Technology*. 1996, Vol. 19.

[9] Pecht, M. Issues Affecting Early Affordable Access to Leading Electronics Technologies by the US Military and Government. *Circuit World.* 1996, Vol. 22, 2.

[10] Kleyner, A., Bender, M. s.l.: *Enhanced reliability prediction method based on merging military standards approach with Manufacturers Warranty data.* Annual Reliability and Maintainability Symposium, 2003.

[11] Derr, James H. Reliability Prediction of Automotive Electronics – How Well Does MIL-HDBK-217-D Stack Up? [Online] 1985. http://papers.sae.org/850531.

[12] O'Conner, P.D.T., *Undue Faith in US MIL-HDBK-217 for Reliability Prediction. IEEE Transaction on Reliability,* Vol. 37, p. 468.

[13] Wong, K.L. What is Wrong with Existing Reliability Prediction Methods? *Quality and Reliability Engineering International.* 1990, Vol. 6.

2

The Need for Reliability Assurance Reference Metrics to Change

2.1 Wear-Out and Technology Obsolescence of Electronics

The vacuum tube or thermionic valve brought the dawn of the age of electronics. At the beginning of the 20th century, vacuum tubes were the main active component in electronics. A vacuum tube (also called an electron tube) is a sealed glass or metal-ceramic enclosure used in electronic circuitry to control the flow of electrons between metal electrodes sealed inside the tubes. A hot filament is used to provide a flow of electrons through a grid with a variable voltage. The filament inside the vacuum, like incandescent light bulbs, would become thinner as the metal evaporated over time. The time for a tube to wear out was dependent on the operating temperature and the quality of the filament and the vacuum. The life of electronics with vacuum tubes was very dependent on its operating temperature.

In the 21st century, most active components have significant life entitlements if they have been correctly manufactured and applied in

Next Generation HALT and HASS: Robust Design of Electronics and Systems, First Edition.
Kirk A. Gray and John J. Paschkewitz.

circuit. Most failures in electronic systems are not due to the intrinsic failure mechanisms of the individual component, but more likely due to an error in application or in interconnections such as poor solder joints and plated through holes (PTH). The vast majority of failures of electronics in about the first 3–5 years are a result of assignable causes somewhere between the design phase and the mass manufacturing process. In cases where a component is misapplied, is overstressed, and fails, it may be assumed to be an intrinsic 'wear-out' without further failure analysis investigation.

2.2 Semiconductor Life Limiting Mechanisms

There are four common intrinsic semiconductor mechanisms in silicon-based ICs that are considered the main physical mechanisms that if the device is used long enough will eventually lead to wear-out failures.

These mechanisms of concern in ICs are electromigration (EM), time dependent dielectric breakdown (TDDB), hot carrier injection (HCI) and negative bias temperature instability (NBTI). There is little evidence of these mechanisms contributing significantly to the unreliability of systems mainly because the time frame for these intrinsic semiconductor mechanisms to reach a failure condition is almost always well beyond the system's technological obsolescence.

The physics of these degradation mechanisms is not completely understood and will inevitably change along with changes in fabrication dimensions, materials and methods. EM, TDDB and NBTI all have positive activation energies, and HCI is actually negative and is inversely proportional to temperature. For accelerated life testing, higher temperatures accelerate EM, TDDB, and NBTI and decelerate HCI.

To compound the difficulty in developing accurate life model derivations for these mechanisms, each mechanism interacts with the voltage acceleration parameters.

[From the NASA paper "Microelectronics Reliability: Physics-of-Failure Based Modeling and Lifetime Evaluation" by Mark White and Joseph B. Bernstein a public domain document]

The failure rate models and acceleration factors for EM, HCI, TDDB, and NBTI are listed below.

1. EM

From the well known Black's equation [1] and Arrhenius model, failure rate of EM can be expressed as:

$$\lambda_{EM}\alpha\left(J\right)^{n}\cdot\exp\left[\frac{-E_{aEM}}{kT}\right] \qquad (2.1)$$

where J is the current density in the interconnect, k is Boltzmann's constant, T is absolute temperature in Kelvin, E_{aEM} is the activation energy, and n is a constant. Both E_{aEM} and n depend on the interconnect metal.

Black's equation model is abstract, not based on a specific physical model, but flexibly describes the failure rate dependence on the temperature, the electrical stress and the specific technology and materials. More adequately described as descriptive as opposed to prescriptive, the values for A, n and Q are found by fitting the model to experimental data. These errors arise from the assumption that the fitting parameters A, E_a and n obtained under accelerated tests are also valid for the life cycle stress of operating conditions, so that they can be directly applied for the life duration extrapolation [2].

Obtaining the real operational stress life cycle environment and the distributions of the conditions across a fielded population is a difficult task. Making assumptions or estimates of the stress life cycle conditions for an electronics system and its distributions is not justifiable if there is no documented or recorded empirical data to support it.

Unfortunately in the field of electronics reliability engineering, relatively simplified assumptions are accepted as valid for deriving the estimated average time to or before failure. It does not cost much to make predictions of a complex system's life entitlement, relative to the time and effort to perform subassembly and systems testing. Yet misleading predictions of system life made based on broad assumptions of critical parameters may result in added costs, such as mechanical cooling, which increases costs, and even potentially reduce a system's reliability through increased complexity and parts.

The experimental data obtained through HTOL (high temperature operating life) for any component will only be valid for the particular component application with the voltage, temperature and use stress factors for the specific circuit it is subjected to in the specific HTOL conditions. In complex systems there may be hundreds or thousands of different metals, voltages, frequencies and temperature conditions for all the semiconductor components used at the system level. Considering the rapid rate of new semiconductor components introduced into the market each year, deriving the values of A, n and Q would take considerable time and expense and could have significant variations between alternative suppliers of the same type of device The total BOM (bill of materials) for an electronics system may contain hundreds or thousands of active semiconductor components of different design and manufacturing vintage. Determining the estimated life from interacting stresses from empirical data is an impractical and almost impossible task. The mixture of new and old designs, materials, fabrication methods and materials in semiconductor devices adds more complexity to the physical degradation models, making the task of collecting valid data:

Recently, copper/low-K dielectric material has been rapidly replacing aluminum alloy/SiO_2-based interconnect. For copper, n has been reported to have values between 1 and 2 [3] and E_{aEM} varies between 0.7 eV and 1.1 eV [4].

In Equation (2.1), current density, J, can be replaced with a voltage function [5]:·

$$J = \frac{C \cdot V_D}{W \cdot H} \cdot f \cdot p \tag{2.2}$$

where C, W, and H are the capacitance, width, and thickness of the interconnect, respectively. f is the frequency and p is the toggling probability; therefore, A_{EM} is also a function of voltage:

$$\lambda_{EM} \, \alpha \, (V_D)^n \cdot exp \left[\frac{-E_{aEM}}{kT} \right] \tag{2.3}$$

The EM acceleration factor is:

$$AF_{EM}^{V_0,T_0;V_A,T_A} = \left(\frac{V_A}{V_0}\right)^n \cdot \exp\left[\frac{E_{aEM} \cdot (T_A - T_0)}{k \cdot T_A \cdot T_0}\right] \qquad (2.4)$$

2. HCI

Based on the empirical HCI voltage lifetime model proposed by Takeda [6] and the Arrhenius relationship, HCI failure rate $AHCI$ can be modeled as:

$$\lambda_{HCI} \, \alpha \, \exp\left[\frac{\gamma_{HCI}}{V_D}\right] \cdot \exp\left[\frac{-E_{aHCI}}{kT}\right] \qquad (2.5)$$

where y_{HCI} is a technology-related constant and E_{aHCI} is the activation energy, which varies between –0.1 eV to –0.2 eV [7].The negative activation energy means HCI becomes worse at low temperature. The HCI acceleration factor is:

$$AF_{HCI}^{V_o T_o;V_A T_A} = \exp\left[\gamma_{HCI} \frac{(V_A - V_O)}{V_A \cdot V_O}\right] \cdot \exp\left[\frac{E_{aHCI} \cdot (T_A - T_O)}{k \cdot T_A \cdot T_O}\right] \qquad (2.6)$$

While the most common equation for the basis of electronics life prediction model is the Arrhenius, in which the reliability is proportional to temperature, the HCI phenomena is inversely proportional to temperature.

3. TDDB

The exponential law for TDDB failure-rate voltage dependence has been widely used in gate oxide reliability characterization and extrapolation. Combining with the Arrhenius relationship for temperature dependence, the TDDB failure rate is:

$$\lambda_{TDDB} \alpha \exp[\gamma_{TDDB} \cdot V_G] \cdot \exp\left[\frac{-E_{aTDDB}}{kT}\right] \qquad (2.7)$$

where y_{TDDB} is a device-related constant and E_{aTDDB} is the activation energy. E_{aTDDB} normally falls in the range of 0.6 eV to 0.9 eV [7]. The

TDDB acceleration factor is:

$$AF_{\text{TDDB}}^{V_o T_o : V_A T_A} = \exp\left[\gamma_{\text{TDDB}} \cdot (V_A - V_O)\right] \cdot \exp\left[\frac{E_{\text{aTDDB}} \cdot (T_A - T_O)}{k \cdot T_A \cdot T_O}\right] \quad (2.8)$$

4. NBTI (Negative Bias Temperature Instability)
Like TDDB, NBTI voltage dependence can also be modeled by the exponential law [8].

Considering the temperature dependence together, the NBTI failure rate is:

$$\lambda_{\text{NBTI}} \propto \exp\left[\gamma_{\text{NBTI}} \cdot V_G\right] \cdot \exp\left[\frac{-E_{\text{aNBTI}}}{kT}\right] \quad (2.9)$$

where γ_{NBTI} is a constant, and E_{aNBTI} is the activation energy, which has been reported to vary from 0.1 eV to 0.84 eV [9]. The NBTI acceleration factor is:

$$AF_{\text{NBTI}}^{V_o T_o : V_A, T_A} = \exp\left[\gamma_{\text{NBTI}} \cdot (V_A - V_O)\right] \exp\left[\frac{E_{\text{aNBTI}} \cdot (T_A - T_O)}{k \cdot T_A \cdot T_O}\right] \quad (2.10)$$

Combined Voltage and Temperature Acceleration Factor
Considering the voltage and temperature acceleration effect together, system acceleration is further complicated by the interplay between voltage and temperature acceleration, as shown above.

Since there is no universal E_{aSYS} and y_{SYS} of multiple failure mechanisms, using AF_T with one activation energy and AF_V with one voltage acceleration parameter for reliability extrapolation is not appropriate. Taking the simulation above as an example, we find out that failure rate estimation using the multiplication model gives an overly optimistic result. The real system failure rate at (50°C, 1.30 V) is **20X** that of the estimated failure rate using the multiplication model with E_{aSYS} and y_{SYS} from high-temperature, high-voltage acceleration testing at (125°C, 1.55 V).

2.2.1 Overly Optimistic and Misleading Estimates

So why do reliability engineers believe in these models with the misleading and overly optimistic estimates?

These four common semiconductor wear-out mechanisms are only related to the semiconductor die itself and not the other mechanisms related to the assembly and encapsulation of the die, such as wire bonding to the lead frame or package delaminating.

To compound the difficulty of reliability predictions, there are so many potential failure mechanisms at each level of component, circuit board assembly and system level interconnections that all the intrinsic critical or dominant failure mechanisms cannot be known.

Relying on the many assumptions that are required to complete the model parameters can lead to more invalid conclusions about life cycle stresses and the impact to the systems life models. The authors of this NASA paper mention these potentially misleading results when multiple mechanisms and the effects of multiple stresses are considered. System life models are many times the basis for traditional accelerated reliability testing, and invalid life models can lead to invalid life test conclusions.

Qualification Based on Failure Mechanism

It is a matter of great complexity to build a system lifetime model to fit all temperatures and voltages if there are multiple failure mechanisms at work.

The conventional extrapolation method using one E_{aSYS} and y_{SYS} tends to give an overly optimistic estimation.

For systems with strict reliability requirements (such as aerospace avionics), more accurate reliability projections are necessary for system design and qualification. Using acceleration parameters obtained at high-temperature, high-voltage acceleration testing cannot be justified because stress conditions tend to accelerate failure mechanisms with high positive activation energy and a larger voltage acceleration parameter, such as TDDB, while EM and HCI failures are more common in field applications. To improve the accuracy of reliability qualification, all failure mechanisms should be considered in the qualification approach.

Table 2.1 Semiconductor wear-out mechanisms
activation energies

Failure mechanism	Voltage acceleration parameter	Activation energy (eV)
EM	2	1.2
HCI	16	−0.2
TDDB	12	0.7
NBTI	6	0.4

Data that is required to accurately model HCI, TDDB, and NBTI is not generally provided. Assumptions must be made about the critical parameters in the models which have no supporting empirical data.

Due to proprietary issues, manufacturer microelectronic device lifetime data is rarely reported in the literature. To reveal the characteristics of temperature and voltage acceleration at the system (component) level, we can perform lifetime simulation by using the models given above. All the activation energies are extracted from published sources.

The activation energies associated with these four failure mechanisms are shown in Table 2.1. The rapid change in materials and dimensions in semiconductor fabrication will modify these wear-out mechanisms, parameters and interactions. As the feature size decreases and the density of devices in an integrated circuit increases, the challenge of controlling parametric variations will increase [9].

2.3 Lack of Root Cause Field Unreliability Data

A major reason for the continued belief in the predictability of failure rates of electronic systems is because the real data of real product failures is locked and guarded as proprietary in most all electronics companies. Root cause failure data from real products is rarely, if ever, released to the public. This lack of data and the evidence that would show that most causes of field failures are not intrinsic to components or assemblies. The lack of data on the real causes of systems failures will continue, and the industry will continue to be misled by perpetuating the belief that reliability of systems can be predicted. The few engineers who see that unreliability in electronics is generally assignable to errors in design, manufacturing or misapplication are not likely to take on the significant challenge of exposing this obvious reason that reliability predications have not been shown to correlate to field failure rates, because the predictions are not based on the real causes of failures.

Because of the long durations required to observe field reliability and because of the fluidity of engineering teams being reorganized for different projects or companies every year, many reliability engineers working at the design stage may never be with a company or project to observe and understand the real causes or return rates once the design is in the field.

Another problem with reliability engineering is the assignment of causes of unreliability in the field. In almost all cases of a product failure in the first several years of use come from overlooked design margins, manufacturing errors and excursions, or customer accidental misuse or abuse. Since field failures can be costly to a manufacturer, no one wants to be held responsible for mistakes that lead to field failures. Many reliability engineers also have a tendency conveniently segment early life latent defect failures arising from manufacturing errors as 'quality problems' and not a 'reliability' problem even though, to a customer, all failures of an electronics product are from poor 'reliability' regardless of which company department is to blame.

Those who find the root causes of verified warranty failures in systems returned from the field in electronics companies know that the vast majority of field failures come from causes that have no connection to the intrinsic failure mechanisms in active or passive components. The vast majority of electronics systems do not fail because of intrinsic physical fatigue damage or chemical degradation which may be considered to have consumed the device's total life entitlement in normal operation and therefore do not have consistent factors to model and predict.

There is no empirical evidence of electronics failure prediction methodology (FPM) correlating to actual electronics failure rates over the many decades that it has been applied. Despite the lack of supporting correlating evidence, FPM and MTBF estimates are still used and referenced for a large number of electronic systems companies. FPM has shown little benefit in producing a reliable product, since there has been no correlation to actual causes of field failure mechanisms or rates of failure. In some cases it has added to program costs, particularly in military procurement, as conservative estimates of MTBF result in stockpiling spare parts that will never be needed.

The Arrhenius equation, used to modify the component failure rates, has been widely misapplied as an acceleration factor (AF) for components. The Arrhenius law of temperature is:

$$AF_T = exp\left[\frac{E_a}{k}\left(\frac{1}{T_1} - \frac{1}{T_2}\right)\right]$$ (2.11)

Where E_a is the activation energy, k is the Boltzmann's constant, and T_1 and T_2 are temperatures in kelvin.

The activation energy has been assumed to be 0.7 eV for use for semiconductor FPM, yet there is no reference to what specific physical mechanism models the activation energy is derived from.

In an electronics system with many types of active semiconductors, applying an average value of 0.7 eV activation energy is an assumption that is not valid across an electronics assembly. Activation energies for the mechanisms shown in Table 2.2 result in an extreme range of AFs. A burn-in process of 40°C above expected use will result in an AF of 3.6 for activation energy of 0.3 eV to an AF of 429 for activation energy of 1.4 eV. Without referencing a specific failure mechanism, the AF derived from the Arrhenius equation is in reality an erroneous equation for electronics reliability prediction.

Semiconductor fabrication methods and materials technologies have changed significantly since 1998, and so has the activation energy of the degradation mechanisms, which makes any Arrhenius-based AF a wild approximated guess and will result in misleading and invalid failure rates of semiconductors.

The belief in the Arrhenius relationship for component failures leads to the conclusions that using the 0.7 eV average value for the activation energy results in the calculation that for every rise of 10°C

Table 2.2 Reported activation energy for silicon semiconductor wear-out mechanisms. Source: Adapted from Jensen and Peterson, 1982 [28]

Silicon semiconductor component and mechanism	Reported E_a
Surface charge accumulation, bipolar	1.0
Surface charge accumulation, MOS	1.2
Slow trapping charge injection	1.3–1.4
Metallization electromigration	0.5–1.2
Corrosion (chemical, galvanic, electrolytic) bonds	0.3–0.6
Intermediate growth Al/Au	1.0

a component's life is approximately reduced by 50% and conversely each decrease of 10°C increases component life by 100%. There is no evidence to support this assertion. Unfortunately, this long-held and invalid assumption may lead to additional assembly and operating costs when fans are added to a design. Fans can introduce other potential causes of failure. They suck in dust and contamination and filters may clog, blocking air flow resulting in system failures due to overheating. Since most fans are mechanical systems with bearings that are subject to material wear, they add other risks such as increasing acoustical noise and bearing wear-out before technological obsolescence.

It is a matter of great complexity to build a system lifetime model to fit all temperatures and voltages if there are multiple failure mechanisms at work. Many if not most of the common failure mechanisms in electronics are accelerated by more than one stress.

Using average values, the critical parameters in these equations to predict wear-out for assemblies with thousands or millions of solder joints of all types, from through hole to ball grid arrays (BGA), does not and will not lead to any reasonable correlation to lifetimes of electronic systems. The values of activation energies for each mechanism require empirical evidence and testing of each relevant failure mechanism. It would be difficult to determine activation energy for the different failure mechanisms to which the Arrhenius equation would be applicable if the manufacturing processes used to manufacture the devices were statistically capable and all relevant parameters were well within the six sigma goals of statistical process control. Needless to say, most manufacturing processes have variations that may or may not be controlled resulting in an even wider distribution of the different wear-out failure mechanisms activation energies and therefore uncertainty in the estimates based on a single average activation energy value.

Many of the potential failure mechanisms in electronics assemblies are related to material bonds, such as cracks in solder or component package delamination, and the Arrhenius based acceleration factors are not relevant for those mechanisms as they are largely driven by thermal cycles and vibration. Several engineering models have been developed for predicting solder joint reliability based on Coffin–Manson acceleration models. A more general approach of applying Coffin–Manson solder fatigue models comes from Morrow's type of fatigue laws, where cycles to failure are given as a function of the cyclic inelastic strain energy density, ΔW_{in}:

$$N = \frac{C}{\Delta W_{in}^{n}} \tag{2.12}$$

Where C is a material constant and n is an exponent that has been found to be in the range of 0.7 to 1.6 for several engineering metals, including soft solders. Assuming a single value for n for all solder joints from through-hole to surface mount j leads to BGAs is not justifiable.

There are many competing physical mechanisms that will degrade over time and lead to wear-out of an active semiconductor component. As noted above, some of the common competing failure mechanisms of active semiconductors over time are time dependent dielectric breakdown (TDDB), electromigration (EM), negative bias temperature instability (NBTI), and hot carrier injection (HCI) mechanisms.

The belief in invalid FPM leads to false predications of increased failure rates, based on temperature, which have led to large producers of IT hardware settting maximum operational environment temperature specifications very conservatively at 35°C. Many parts of the world that do not have air-conditioning will exceed 35°C. And this belief also prevents some IT suppliers bidding for millions of dollars of potential sales because the warranty would be invalid for use conditions over 35°C. Reliability engineering at these IT systems suppliers would not support an operation specification increase of 5°C because the failure rate would increase by 50%, according to traditional electronics prediction methodology. This was despite the fact that many of the same systems were demonstrated to operate reliably for thousands of hours at 65°C [10]. Hundreds of millions of dollars in sales of IT client hardware may have been lost due to the continuing belief in the mythology of traditional reliability predictions.

2.4 Predicting Reliability

Prediction of future events has been a universal desire of humanity from its beginning. Everyone desires to know the uncertain future. Some future events, such as the Earth's days and seasons have a high probability of occurrence due to our knowledge of nature. But prediction of the life entitlement of electronics is similar to weather predictions. As Albert Einstein said about the weather, 'When the number of factors coming into play in a phenomenological complex is too large, scientific method in most cases fails. One need only think of the weather, in which case the prediction even for a few days ahead is impossible.' Knowing where, when and how powerful a hurricane or tornado may hit land more than a few days in advance is still beyond the realm of predictability, even though accurate predictions of hurricanes and other destructive weather events would save millions of human lives and millions in destruction of homes and buildings.

Media coverage of a major failure of electronic systems in recent times illustrates the difficulty and failure of system modeling and reliability predictions.

> 'The assumptions used to certify the battery must be reconsidered,' said NTSB Chairwoman Deborah Hersman. 'The design and certification assessment, and the assumptions that were made, were not borne out by what we saw in flight experience.' 'The 787 fleet has accumulated less than 100,000 flight hours,' she said. 'Yet there have now been two battery events resulting in smoke less than two weeks apart on two different aircraft.'[1]

---

[1] http://seattletimes.com/html/businesstechnology/2020307773_ntsb787xml.html The Seattle Times, 7 February 2013.

Figure 2.1   Burned battery assembly after suffering a thermal runaway. Source: NTSB, 2013

In the Boeing 787 Dreamliner's first year of service, at least four aircraft suffered from electrical system problems stemming from its lithium-ion batteries. Despite Boeing's analysis and testing (it is not known if they used HALT methods), they were unable to discover the latent weakness in the battery system that has caused the thermal runaway events resulting in heavily burned battery shown in Figure 2.1.

The root cause of the battery thermal runaway has never been determined. The Federal Aviation Administration decided on 19 April 2013, to allow the US Dreamliner to return to service after changes were made to their battery systems to contain battery fires better. These changes did not prevent the battery system failures, and the battery system was still having thermal runaway events as recently as January 2014, when a battery in a Japanese Airlines 787 emitted smoke from its exhaust and was partially melted while the aircraft was undergoing pre-flight maintenance [11].

The battery failures are an example of how difficult it is to predict reliability and how costly latent defects can be when found in service. Many times it is argued that using HALT to stress a system to operational failure, and sometimes destruction, is cost prohibitive. Although the cost of these battery failures has not been reported, the costs of performing tests that may destroy several expensive new battery systems in a HALT evaluation would be small in comparison to the cumulative monetary losses Boeing and its customers have incurred with the grounding of the aircraft.

There has not been any published evidence that reliability predictions have ever correlated to the field reliability data.

The following section explains the invalidity and misdirection of traditional reliability prediction for electronics. The paper from which the section is taken was first presented at the 2013 Annual Reliability and Maintainability Symposium. This U.S. Government paper reinforces why a change from probabilistic predictions to deterministic reliability analysis is a better solution.

2.5 Reliability Predictions – Continued Reliance on a Misleading Approach[2]

Reliability prediction methodologies, especially those centered on Military Handbook (MIL-HDBK) 217 and its progeny are highly controversial in their application. Reliability predictions in the design and operation of military applications have been used since the 1950s. Various textbooks, articles and workshops have provided insight into the pros and cons of these prediction methodologies. Recent research shows that these methods have produced highly inaccurate results when compared to actual test data for a number of military programs. These inaccuracies promote poor programmatic and design decisions, and often lead to reliability problems later in development. Major reasons for handbook prediction inaccuracies include, but are not limited to the following:

1. The handbook database cannot keep pace with the rapid advances in the electronic industry.
2. Only a small portion of the overall system failure rate is addressed.
3. Prediction methodologies rely solely on simple heuristics rather than considering sound engineering design principles.

---

[2] From *Reliability Predictions – Continued Reliance on a Misleading Approach* by Christopher Jais, US Army Material Systems Analysis Activity; Benjamin Werner, US Army Material Systems Analysis Activity; and Diganta Das, Center for Advanced Life Cycle Engineering, University of Maryland College Park.

Rather than relying on inaccurate handbook methodologies, a reliability assessment methodology is recommended. The reliability assessment methodology includes utilizing reliability data from comparable systems, historical test data and leveraging subject-matter-expert input. System developers then apply fault-tree analysis (or similar analyses) to identify weaknesses in the system design. The elements of the fault tree are assessed against well-defined criteria to determine where additional testing and design for reliability efforts are needed. This assessment methodology becomes a tool for reliability engineers, and ultimately program managers, to manage the risk of their reliability program early in the design phase when information is limited.

2.5.1 Introduction

The use of reliability predictions in military applications produces misleading and inaccurate results [14]. The National Academy of Sciences, along with lessons learned from the US Department of Defense (DoD) over the past decade, suggests several reasons why military systems fail to achieve their reliability requirements. These reasons include a 'reliance on predictions instead of conducting engineering design analysis.' [15]

Reliability predictions represent a single number that attempts to describe a complex system through the estimation of its failure rate. Although predictions can be a valuable tool in the design process, they are often improperly developed, misreported and/or misinterpreted. A main reason for this problem is the use of MIL-HDBK-217 and associated methods. These methods include any handbooks or commercial applications based on MIL-HDBK-217 (e.g. Telcordia/Bellcore, HRD, PRISM, 217Plus, etc.). MIL-HDBK-217 uses historical data of electronic systems to determine a constant failure rate of electronic parts. The associated part prediction is a function of a generic failure rate and a series of adjustment factors. The final system-level prediction assumes a series structure and is a summation of the individual electronic parts. Because of the technical limitations associated with the prediction documents, as discussed here, the handbook results have no connection to real product reliability and can in fact promote poor reliability practices and reliability decisions.

Here we discuss the limitations of the MIL-HDBK-217 methodology, its continued misuse in military applications and an alternative method for assessing reliability early in system development that provides more valuable insight for both the system developer and the customer.

2.5.2 Prediction History

Reliability prediction approaches started soon after World War II with the formation of several ad hoc reliability groups. The desire of these groups was to standardize requirements and improve the reliability of increasingly complex electronic components. The original version of MIL-HDBK-217 was published in April 1962 by the US Navy. The first revision, MIL-HDBK-217A, occurred in December 1965. MIL-HDBK-217A became the standard for reliability predictions. The main reason for its ascension was that it was often cited in contractual documents [16].

In 1974 the responsibility for preparing MIL-HDBK-217 was transferred to RADC, under the preparing activity of the US Air Force. They published Revision B and addressed the rapidly changing technology. They also incorporated overly simplified versions of the RCA models, which are still in the handbook nearly 40 years later [14].

As electronics grew more complex MIL-HDBK-217B received several changes, eventually leading to MIL-HDBK-217 Revision C. The 1980s brought about Revisions D and E of MIL-HDBK-217 attempting to keeping pace with the changes in technology. The 1980s also brought several reliability prediction models unique to selected industries. Examples of these include the Society of Automotive Engineers Reliability Standards Committee and Bell Communications Research (now Telcordia). These industries, among others, based their prediction techniques on the MIL-HDBK-217 models.

In December 1991, RADC (now renamed Rome Laboratory) released MIL-HDBK-217 Revision F. In 1994, the former US Secretary of Defense, Dr William J. Perry, announced the reduction of reliance on military specifications and standards and encouraged the development of commercial standards that could be used by the military, in his memorandum, 'Specifications & Standards – A New Way of Doing Business.' In 1995, the redistribution of MIL-HDBK-217F contained the following notice, 'This handbook is for guidance only. This handbook shall not be cited as a requirement. If it is, the contractor does not have to comply.' The following year the Assistant Secretary of the Army for Research, Development and Acquisition, Gilbert F. Decker, declared that MIL-HDBK-217 was not to appear in any Army request for proposal acquisition requirements [17].

Since 1995, there has been no update to MIL-HDBK-217. However, there have been efforts by an industry working group to update the standard. The working group, which was led by the Naval Surface Warfare Center (NSWC) at Crane, IN and consisted of government and private industry personnel, developed a three-phase plan for revisions. All three phases were planned to be completed by December 2011 [18]. However, the effort to acquire appropriate data, and differences in opinion on the methodologies to be incorporated, has led to significant delays with no revisions published.

2.5.3 Technical Limitations

Reliability predictions can be useful when determining early-on reliability allocations or forecasting life-cycle costs. However, the technical limitations of MIL-HDBK-217 methodologies misrepresent a system's true reliability metric (i.e. reliability mean time between failure, mean miles between system abort, etc.). Technical limitations of MIL-HDBK-217 have been a topic of debate since its development in the 1960s, with copious research examining its strengths and weaknesses. Four major limitations of these methodologies that impact DoD system design and development are discussed in the following sections.

2.5.4 Keeping Handbooks Up-to-Date

MIL-HDBK-217 has not been updated since 1995. When a developer uses it for predicting a system's reliability today, well over 15 years of technology is not included. Prior to 1995 there were only six major updates since its original release in 1962. During this time new devices were not covered for approximately five to eight years, penalizing system developers for utilizing new technology. Revisions also failed to update connector models for over 35 years. Handbook models also require historical field data. This data is acquired from a variety of sources, over different periods of time, and under various field conditions. No standard for verification or statistical control of this data exists. The handbooks do not supply information regarding any of these factors.

Given these limitations the handbook databases cannot keep pace with the rapid advances in electronics technology and products. Any plans to simply update the database and models would exclude any emerging technology.

2.5.4.1 System Failure Rate

Reliability estimates of MIL-HDBK-217 methodologies assume a constant failure rate. However, electronic component failure rates can vary depending on many factors including the usage conditions and the remaining life of the component. Instead of assuming the system or the component to be a black box, a better understanding of how and why components fail can be obtained by studying the physics of failure [19]. For example, for power electronic modules and insulated gate bipolar transistors (IGBTs), wire bond failure and die attach failure have been found to be the two most dominant and critical failure mechanisms [20]. These mechanisms could induce failures in the package, depending on the usage and loading conditions and thus cannot be represented by a constant failure rate. However, the mechanisms and their associated time to failure can be characterized by well-established models and equations.

While power electronics is specifically addressed above, Pecht et al. [21] have discussed failure mechanisms found in other applications in the field. Similarly, based on the field returns, the manufacturers can identify the dominant failure mechanisms, identify the associated models and use them to estimate the lifetime of components being used in a particular application, under certain conditions.

Even if MIL-HDBK-217 methodologies accurately depicted electronic parts failure rates, they would only account for a small portion of the overall system's failure rate, as depicted in Table 2.3. DoD systems follow the same trend. Figure 2.2 displays the chargeability (determined cause for a failure) for a DoD network and aviation system. Hardware failures account for only 7% and 47% of the overall system's failure rate, respectively. It should be cautioned that the hardware failures represent both mechanical and electronic failures and therefore the failure rate due to electronic components may be even smaller. System predictions should not only account for electronic components, but must also factor in failure models due to design, manufacturing, wear-out, software and external factors (crew/maintainers).

2.5.4.2 Critical Design Factors

Prediction methodologies do not consider sound engineering design principles. For example, handbook predictions for a circuit card are not affected by how the device is mounted and supported, the natural frequency of the board or where the largest

Table 2.3 Causes of Failure

Category of Failure	Study 1 [16]	Study 2 [14]
Parts	22%	16%
Design Related	9%	21%
Manufacturing Related	15%	18%
Externally Induced	12%	–
No Defect Found	20%	28%
Other (wear out, software, management, etc)	22%	17%

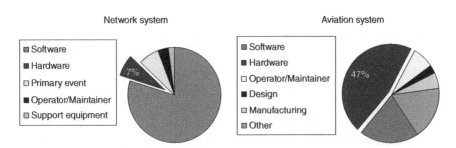

Figure 2.2 Chargeability of failures based upon test data

deflections are located in relation to the components. They do not consider the impact of temperature cycling, humidity cycling, vibrations or mechanical shock throughout the components' life cycles. The life cycle of a product consists of manufacturing, storage, handling, operating and non-operating conditions. The life-cycle loads, either individually or in various combinations, may lead to performance or physical degradation of the product [13]. Extensive research shows the effect of thermal aging and thermal cycling. This research demonstrates the need to account for multiple deployments with sequential thermal stresses and uncontrolled thermal environments [22]. Handbook methodologies overemphasize steady-state temperature and voltage as operational stresses and do not take into account any of these engineering design decisions. For example, the use of MIL-HDBK-217 methods led to poor design decisions on the F-22 advanced tactical fighter and the Comanche helicopter [23]. In both cases the designs indicated the need for significantly lower temperatures of the avionics components. The resulting temperature cycling created unique failure mechanisms that ultimately impacted the cost and schedule of both programs.

2.5.4.3 Insight into How or Why a Failure Occurs

Practitioners use handbook predictions as a design tool. The pitfall of using predictions is that the methodology does not give insight into the actual causes of failure since the cause–effect relationships impacting reliability are not captured. Therefore, the developers cannot implement the appropriate corrective action

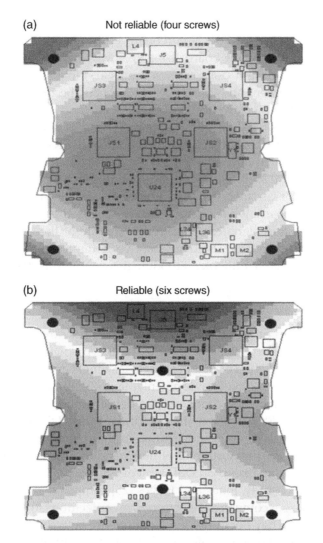

Figure 2.3   Comparison of vibration displacement

or mitigation plan. Handbooks simply sum the failure rates from the total parts on a given component. An example of this can be seen by examining the vibration displacement for a circuit board. Although the components and their placement on the two circuit boards in Figure 2.3 are the same, the reliabilities are significantly different. In this example circuit board (a) is a four-screw configuration versus circuit board (b), a six-screw configuration. The difference in design (four screws versus six) impacts the vibration displacement and consequently impacts the reliability. The addition of two screws to the design significantly increased the circuit board's reliability. However, both designs would have the exact same reliability prediction using MIL-HDBK-217.

The placement of components is another crucial design consideration. Figure 2.4 considers the placement of a surface mount network resistor. Circuit board (a) places the resistor in a high vibration area of the board, while circuit board (b) moves the resistor to the outer edge and significantly increases the life of the component and the circuit board. These are just two examples of design considerations that handbook methodologies do not consider.

2.5.5 Technical Studies – Past and Present

Since the inception of MIL-HDBK-217 there have been several studies examining the inaccuracies of the prediction numbers. Cushing et al. [24] explored the Single Channel Ground Air Radio Set (SINCGARS) Non-Developmental Item (NDI) Candidate Test. In their research they compared the demonstrated test MTBF of nine SINCGARS vendors to their predicted MIL-HDBK-217 MTBF. Table 2.4 displays the results shown in that paper. These results were one of the first examples of how handbook predictions produce misleading results on DoD systems.

In another study, Jones and Hayes [25] compared circuit board field data from commercial electronics manufacturers to handbook predictions. They not only found a difference between the prediction and the field failure rate, but also found significant differences between handbook methodologies. Figure 2.5 shows the results discussed in the paper.

These are just two examples of the previous work done to compare handbook predictions with demonstrated reliability estimates. The literature is scattered with additional examples citing the significant differences between predicted and demonstrated failure rates for components and parts.

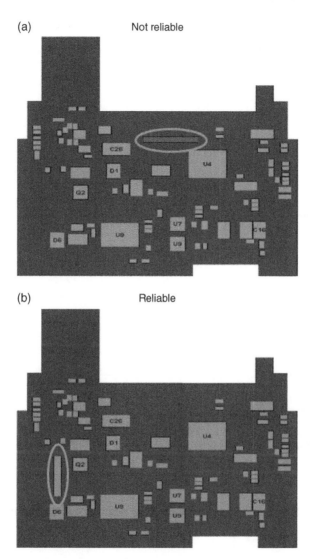

Figure 2.4 Comparison of vibration response and resistor location

Despite these results and the documented technical limitations of predictions, there are still several reports that support the use of the current handbook methodologies. Brown [26] used the Modular Airborne Radar program (a US Air Force system) to compare field

Table 2.4   Results of the 1987 SINCGARS NDI candidate test

Vendor	MIL-HDBK-217 MTBF (hours)	Actual test MTBF (hours)
A	7247	1160
B	5765	74
C	3500	624
D	2500	2174
E	2500	51
F	2000	1056
G	1600	3612
H	1400	98
I	1250	472

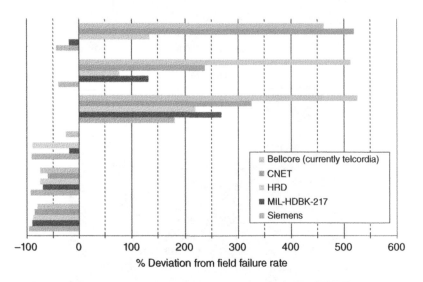

Figure 2.5   Comparison of various handbook methodologies

data from plastic encapsulated microcircuits to two prediction tools (MIL-HDBK-217 and a commercial tool based on MIL-HDBK-217). Initial findings revealed that the predictions were optimistic in comparison to the observed field performance. Further evaluation showed that modifying the default values of the model improved the accuracy of the prediction. She also noted that the use of experience data (field data) proved valuable in refining the prediction results. In addition to this, Smith and Womack [27] compared a commercial prediction tool (based on MIL-HDBK-217 methodologies) to actual observed field failure rate for three military electronic units. The initial results showed that the predictions were approximately one-half of the observed field failure rate. This was in contrast to an earlier study by TRW Automotive which showed the predicted failure rates were approximately twice the actual field values. Just as in Brown's study, they found that experience data aided in refining their prediction estimates.

The US Army Material Systems Analysis Activity (AMSAA) recently surveyed various agencies throughout DoD requesting system level predictions and demonstrated results (either from testing or fielding). When compiling the data, only those systems whose predictions where solely developed using MIL-HDBK-217 or its progeny were examined. If the prediction was a combination of field data and predictions it was excluded from the final analysis. Figure 2.6 displays the results of the survey. In total the survey explored 15 systems. One missile system is excluded from Figure 2.6 for graphical purposes (the only system without a mean time between failure metric). These systems represent a variety of platforms to include communications devices, networks command and control, ground systems, missile launchers, air command and control, aviation warning and aviation training systems. The ratio of predictions to demonstrated values ranges from 1.2:1 to 218:1. This shows that original contractor predictions for DoD systems greatly exceed the demonstrated results. In addition, statistical analysis of the data using Spearman's rank order correlation coefficient show that MIL-HDBK-217 based predictions cannot support comparisons between systems. This data demonstrates the inaccuracies of predicted reliability using handbooks to demonstrated results. It should also be noted that these predictions could lead to improper programmatic decisions impacting reliability (minimizing growth testing, design for reliability (DfR) activities, etc.).

These results demonstrate the misuse of predictions in the DoD with the same consequences (unreliable systems with high operating and sustainment costs) as documented in the DoD *Guide*

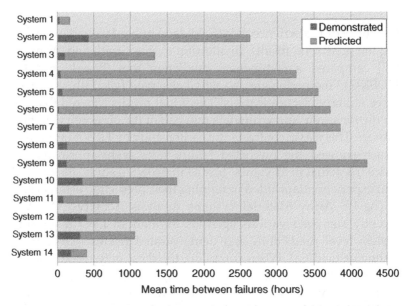

Figure 2.6 Comparison of predicted versus demonstrated values for DoD systems

for Achieving Reliability, Availability, and Maintainability. This begs the question of why, despite its known technical inadequacies and misleading results, is MIL-HDBK-217 still being used in Department of Defense Acquisition? There are several potential answers, but the most prominent is that despite its shortcomings, system developers are familiar with MIL-HDBK-217and its progeny. It allows them a 'one size fits all' tool that does not require additional analysis or engineering expertise. The lack of direction in contractual language also leaves government agencies open to its use.

2.5.6 Reliability Assessment

When system developers are asked to provide a reliability prediction as part of the contract there are two issues:

1. The source of the prediction
2. The method for the prediction

Based upon data from the Naval Surface Warfare Center Crane Division, approximately 50% of reliability predictions have no traceable source. The 23% that had a traceable prediction turned to MIL-HDBK-217 or its progeny 44% of the time despite the limitations and inaccurate results (as demonstrated in the previous sections).

The purpose of predictions is more than just a need for a 'reliability number'. It should be cautioned that simply updating MIL-HDBK-217 based upon current technology does not alleviate the underlying fundamental technical limitations addressed in the earlier sections. Predictions should provide design information on failure modes and mechanisms that can be used to mitigate the risk of failure by implementing design changes.[3]

2.5.7 Efforts to Improve Tools and Their Limitations

The limitations of MIL-HNBK-217 have been well documented as the previous sections has shown. Since the last revision of 217, efforts have been made to improve the reliability prediction tools for electronic and electromechanical systems.

The efforts to improve prediction tools have resulted in several reference documents and guides including:

- UTE C80-810
- Siemens SN-29500
- IEC TR62380
- British Telecom HRD5
- FIDES.

Although these efforts have addressed some of the weaknesses of MIL-HDBK-217, the fundamental lack of application data and root cause failure mechanisms, as well as insistence on considering only the

---

[3] End of section "Reliability Predictions—Continued Reliance on a Misleading Approach".

constant failure rate portion of the life cycle in making predictions are still major limitations. FIDES attempts to include stresses and field failure data, but the databases required to do so are just not commonly available in most industries outside of aircraft and defense equipment. The resources to collect and analyze such data are also beyond the capability of most companies.

2.6 Stress–Strength Diagram and Electronics Capability

It is time for a new frame of reference, the new paradigm that drives the use of the HALT and HASS philosophies, for the reliability assurance measurement of electronics reliability. In the new frame of reference, stress limits will be used for confirming and comparing the capability of electronic systems to meet their reliability requirements when designed. During the production phase the reliability entitlement of the design will be dependent on the capability of the manufacturing process to produce products with the same operational limits and stress margins.

The new orientation for reliability assurance could and should be based on the stress–strength interference perspective and the stress margin capability of the technology. Electronics technology is constantly changing and with the time frames required for modeling new technologies it is like trying to hit a fast moving target with a very slow bullet. There have been changes in materials and fabrication at all levels of electronics system and component designs and manufacturing. But whereas it may take months or years to simulate and model the anticipated life cycle environmental profile (LCEP), stressing a system to find operational limits is relatively quick, taking only days or weeks at most to determine limits and the variation between limits. Operational stress limits between samples during HALT evaluation can be a good indicator of the distribution of strength margins in the design. The wider the distribution in strength margins, the higher the risks of failure.

The reference points from which to determine whether a product has sufficient reliability should come from comparisons of operational and destruct limits under stress and from what is known of the engineering physics that will eventually lead to the inherent wear-out of the device

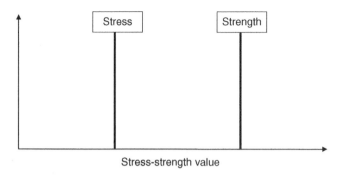

Figure 2.7 The stress–strength diagram for reliability

or system. Not all weaknesses in an electronic system will necessarily result in failures. The stress and strength curves in stress–strength diagrams are probability density functions (PDF) that may be more accurately determined at the finite element analysis (FEA) level. It becomes much more difficult to know the distributions of a total system's strength from computer-aided modeling because some weak units in benign environments may never fail and we have little control over what environment an individual system may be subjected to.

The new metric and relationship to reliability is illustrated in a stress–strength graph as shown in Figure 2.7. This graphic shows the relationship between a system's strength and the stress or load it is subjected to. As long as the cumulative field stresses are less than the strength of the system, no failures occur. Anywhere the stress and strength are equal, failures will occur. It is relevant for any physical structure, from bridges to electronic systems.

For any electronics produced in volume there will be lot-to-lot variations in strength about a mean design value. Some units will have a higher strength and some lower as a result of the manufacturing process. Manufacturing variations should be minimized with using statistical process control (SPC) methods so that the strength of the product is as uniform as possible. Field stresses are not generally controlled and there will be a much wider and often unknown distribution of stress conditions. Some units will have a benign use environment and some will have a much more stressful use conditions. When the product is used in the field environment, the stresses that the product is subjected to will induce cumulative fatigue damage that decreases its strength

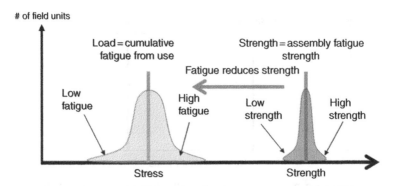

Figure 2.8 The stress–strength diagram and the effect of fatigue damage

over time. The decrease in strength moves the mean of the strength curve to left and towards the stress curve as shown in Figure 2.8.

The intersection between the stress and strength curves for any electronics system is difficult to assess and quantify. The strength distribution is dependent on the consistency and capability of the manufacturing process and will change during its production cycle. The distribution of the cumulative fatigue created from the end use stress conditions is much less controlled and a system may be used in an environment for which the producer had not considered or intended it for.

In production and distribution of electronic systems there will be variations in the strength when manufactured and in the stress conditions it will be subjected to during its use. It is important to remember that PDF curves of stress and strength distributions are probability 'clouds' that are dynamic and have no discrete boundaries. The two curves are probably not normal or symmetrical at any point in time. It is difficult to control or restrict the customer's use in field stress conditions, as the customer may not be aware of, or follow, the recommended or specific operation rating conditions.

For electronics, the environmental specifications are probably not known to most users. In many cases it would restrict the use in normal conditions if it were strictly adhered to. Most consumers of electronics are not aware of the published operating environmental specifications for common consumer devices we all use. For instance, most cell phones are not waterproof, and the user would likely understand that most cell phones are not water submersible. If it were accidently

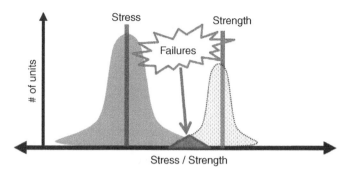

Figure 2.9 The intersection of the stress and strength curves resulting in failure PDFs

dropped into water, most users would understand and accept that it may not operate after that and would take the blame for abusing it.

With thermal and mechanical shock conditions it is not as clear cut. For instance, the Apple iPhone® 5 has a temperature environmental operation specification of 35°C [12] so if it is at human body temperature (37°C) or outdoors on a day of over 35°C when it is turned on, the manufacturer's operating specifications are being exceeded. The non-operating storage specification for the iPhone 5 is 45°C. If it is left in an automobile on a hot summer day, the temperature could exceed 60°C, which is 15°C above the manufacturers storage specification temperature. If a cell phone with the same specifications were left for hours in a automobile at these conditions then failed after being allowed to cool below 35°C, would the user feel they were to blame causing the phones failure or would they place responsibility for failure on the manufacturer because of a less than robust design?

In the stress–strength graph in Figure 2.9, anywhere the load to a system exceeds the system's strength is where the two curves overlap and the area under the curve is a PDF of the probability of failures occurring. This relationship is true for bridges and buildings as well as for electronic systems. If the weight of vehicles on a bridge exceeds the strength of the bridge's structure it will structurally fail. If an electronics system is subjected to stresses that exceed the strength of the circuit elements, it has a higher probability of failure. The intersection of the stress and strength curves results in a normal PDF for failures.

This relationship between stress–strength and failures correlates with our common understanding that the greater the inherent

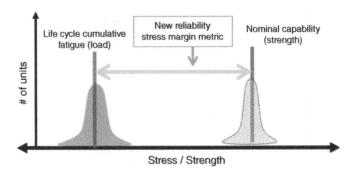

Figure 2.10 Reliability margin in the stress–strength diagram

mechanical strength a system has relative to the mechanical fatigue damage or wear and tear of environmental stress and use conditions, the higher the probability of it not failing in time. We can refer to the space between the mean strength and the mean stress as the reliability margin shown in Figure 2.10.

It is similar to a safety margin or guardbands in a component or system design, except that it is based on the strength of the electronic assemblies under stress, and is not a statistically derived margin based on probable risks.

2.7 Testing to Discover Reliability Risks

There is no single environmental stress test that can precipitate and detect every latent defect, and HALT is no exception. HALT is a very comprehensive test of the robustness of a system or assembly but it does not always discover weak links that need to be mitigated to ensure reliability throughout the product's life cycle. Since HALT uses all the empirical strength of the product under stress to find weaknesses, it does find more opportunities for improving the strength of the design faster than most other testing strategies. After a company has used the HALT methodology it becomes experienced with how to develop a robust assembly from what has been discovered in its previous designs. For companies experienced in using HALT for reliability development, a HALT on a new product can demonstrate that its design is as strong and robust as possible with standard materials and manufacturing methods, otherwise known as the 'fundamental limit of technology' (FLT). If a product is found to have the capability of the FLT in HALT, no changes are needed and the HALT results can be used to establish the HASS stress regime if HASS is used.

A significant difference in field simulation or accelerated life tests (ALT) to quantify field lifetimes is that the stress levels used may be only 10–20% above expected LCEP, and the time and number of test samples required can be very large. Also, depending on the test validation time requirements, ALT may not find a failure point or an empirical limit reference point, and therefore not find any weaknesses or causes of failure that can be evaluated for possible reliability improvement.

HALT is faster and uses fewer samples to gain the most strength data in the shortest time. Typical HALT of a sample can be performed in less than one week using only three to five samples. HALT results in variable stress limit data which, if there is a large deviation between samples, may be an indicator of poor manufacturing process control. Stress limits can also be used to compare benchmarked stress margins found in previous HALT work on previous products. HALT provides benchmarks for determining and achieving the same robustness for the new designs.

The following information in section 2.8 is from Fred Schenkelberg an experienced Reliability Engineering Consultant. It provides the mathematical explanation for using the stress–strength frame of reference to determine probability of reliability if the distributions of PDFs of stress and strength are assumed to be normal.

2.8 Stress–Strength Normal Assumption

Fred Schenkelberg

Ideally, in every design of every component the stress–strength relationship looks like Figure 2.11. The stress is well below the strength.

This implies that there is very little chance of failure due to the element being overstressed. Also, ideally, we fully characterize all stresses and all strengths for each element of a product. This is generally difficult to accomplish and it is rarely done to that extent. In practice we narrow down the list of critical parts and then perform the stress–strength calculations.

There are occasions where there is a definite possibility that some of the elements will experience the chance of stresses that are higher than that element's ability to survive. It is this intersection between the two

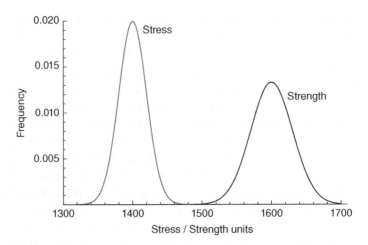

Figure 2.11 The stress–strength curves in a reliable system

curves that provides the probability of failure as shown in Figure 2.12. To calculate that area is to define the double integral that solves for the probability of the stress being higher than the strength. Think of a small normal curve under the overlapping curves.

It is important to note that the area under both curves is not shaded, which would overstate the probability of failure. It can be shown that when both stress and strength are normally distributed the probability of failure is itself a normal curve.

2.8.1 Notation

When we fully characterize the stress and strength, we can often use a probability distribution to describe the location and variation of the values.

1. Probability of failure, $pf = P(Y < X)$
2. Strength, Y is a random variable with mean μ_y and standard deviation σ_y
3. Stress, X is a random variable with mean μ_x and standard deviation σ_x
4. Safety factor $= \mu_y / \mu_x$
5. Safety margin $= \mu_y - \mu_x$

The random variables can be described by any distribution.

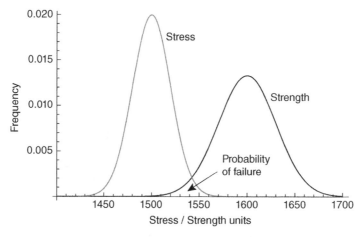

Figure 2.12   The stress–strength curves overlap results in failure PDF

2.8.2 Three Cases

Given that we may have incomplete information or only estimates for either stress or strength, one of three cases may apply for the stress–strength calculation. The first two permit the calculation of the probability of failure directly by using the probability density function, PDF. The last case may require using some calculus.

1. Fixed and known strength, but random variable for stress, X (Figure 2.13).

$$Pf = \int_{y}^{\infty} f_x(x)dx \qquad (2.13)$$

   Failure occurs if stress exceeds known strength. It is the area under the stress distribution to the right of the known strength value.

2. Fixed and known stress, but random variable for strength, Y (Figure 2.14).

$$Pf = \int_{0}^{x} f_y(y)dy \qquad (2.14)$$

   Failure occurs if the strength falls below the known stress. It is the area under the strength distribution to the left of the known stress.

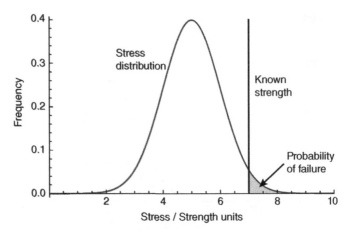

Figure 2.13 Fixed and known strength, but random variable for stress, *X*

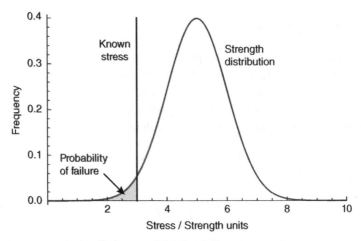

Figure 2.14 Fixed and known stress but random variable for strength, *Y*

3. Both stress and strength are random variables (Figure 2.15).

Failure occurs when the stress is greater than the strength. It is the probability represented by the area under the two curves. It is a distribution of its own, which we'll approximate in the next section.

$$Pf = \int_0^\infty \int_0^x f_x(x) f_y(y) dy dx \qquad (2.15)$$

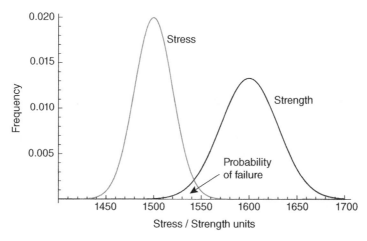

Figure 2.15   Both stress and strength are random variables

2.8.3 Two Normal Distributions

A special case that may apply in your situation is when both stress and strength reasonably fit a normal distribution. When this occurs, we do not have to solve the double integral. It is the difference between the two distributions that is of interest, $D = Y - X$. This difference has the properties of a normal distribution with

$$\mu_D = \mu_y - \mu_x \qquad (2.16)$$

This difference in means is also called the safety margin.

$$\sigma_D^2 = \sigma_y^2 + \sigma_x^2 \qquad (2.17)$$

This permits the direct calculation of the probability of failure if these two curves are known.

2.8.4 Probability of Failure Calculation

Given the stress distribution with a mean, μ_x, of 1500 and standard deviation, σ_x, of 20, and given a strength distribution with a mean, μ_y, of 1600 and standard deviation, σ_y, of 30, determine the probability of failure.

$$Pf = \Phi\left(-\frac{1600-1500}{\sqrt{30^2+20^2}}\right) = \Phi\left(-\frac{100}{\sqrt{1300}}\right) = \Phi(-2.77) = 0.0028 \quad (2.18)$$

2.9 A Major Challenge – Distributions Data

All of the previous statistical analysis requires knowing the distribution of the environmental stresses and the strength of the devices or system. It may be applicable in a limited number of cases, but in the vast majority of electronic systems, the distributions of the LCEP of a significant number of samples for a period of years of the product use will be difficult if not impossible to obtain. The distributions are probabilistic and for many electronic devices, especially portable consumer electronics such as portable PCs, tablets and smartphones, the distributions of stresses extends past destruct levels as when a user accidently runs over it with an automobile or it falls from a high rise building. The consumer is the final judge of what should and should not be survivable stresses for any product, and that judgement is a variable with a wide distribution.

Components, materials and manufacturing processes all contribute to total strength distribution, and all will have variations independently about a nominal value throughout the manufacturing product cycle. For high volume products as in consumer or IT hardware, there are typically multiple suppliers of components to ensure that production volume will meet the market demands, and mitigate the risk of a single supplier causing production to stop if a part is unavailable.

Although the components may have the same advertised parametric specifications on the components data sheet, they may have different distributions and margins on critical parameters. Although a worst case analysis (WCA) and the effect of individual components' parametric variations on the system can be verified, modeling the effect of multiple components, the combinations of first and second sources of components, along with the potential parametric variations introduced by manufacturing (which may not even be known), can rapidly become a very complex computation even for a small number of components.

On the other side of the stress–strength analysis equation is the determining the cumulative stress for a system and the distribution of stress life cycles for the product field population.

Monitoring of stress conditions has – for the purpose of assessing the need for maintenance or to provide an alert for the probability of impending failure conditions or health assessments – been routinely utilized in mechanical systems, civil structures and aircraft. For instance, in rotating machinery the vibration frequency spectrum monitored near bearings or the shaft will change as the bearings or bushings wear. Monitoring the change in vibration spectrum during start-up or increased loading can provide prognostics for maintenance or replacement of mechanical components such as bearings or motors before failure. These methods are referred to as condition monitoring. Similar monitoring and electronics parametric shifts as electronic and electromechanical systems fatigue and wear are being proposed for electronic systems. Using sensors or parametric data to record environmental stress conditions for electronics is proposed for use in prognostic and health management (PHM) of electronics. PHM is based on being able to measure and model an electronic system's parametric changes that are the result of aging degradation to estimate the remaining useful life (RUL).

It is difficult to determine in advance of design and deployment of a new electronic system the locations, precision or frequency of stresses needing to be measured to make an accurate assessment of the reliability by monitoring the degradation and fatigue damage of the system. The total life cycle stress includes manufacturing, shipping, storage, handling, non-operating and operating conditions. Methods of measurement of the relevant stresses and measuring stress events and environment conditions in situ has been utilized in some portable IT hardware systems [13]. Hopefully verification on how well they will be to able to determine a product's RUL will be reported in the future.

2.10 HALT Maximizes the Design's Mean Strength

The overall strength of an electronic or electromechanical system is a sum of the distributions of materials and individual devices used to create the system. There are distributions starting with the strength of components, the raw PWB circuit board, solders, interconnections and connectors and attachment hardware.

At each level of assembly variation occurs during the manufacturing period, and the timings of the variations in distributions are not correlated to each other. A graphical illustration of the subsystems strength

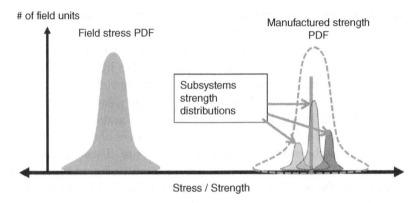

Figure 2.16 The distributions of a system's strength is a sum of individual components and subsystems, each with its own distribution

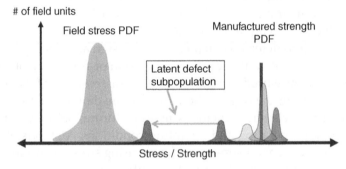

Figure 2.17 A latent defect subpopulation resulting from a manufacturing process excursion

distributions that comprise the total system strength distribution is shown in Figure 2.16. At any time, multiple shifts in the distributions of strength can result in a significantly weak subpopulation that will intersect with a distribution product population that is subjected to the highest life cycle stresses. A graphical illustration of the result is shown in Figure 2.17.

The difficulty in determining the probability of failures using the mathematical relationships of the previously shown examples of stress and strength distribution curves is that the distributions of life cycle environmental stresses for a large population of the same product are difficult to assess, and cannot be assumed to be normal. Analytical process tools such as statistical process control can be utilized to reduce the manufacturing variances if used properly and if the parameters that are monitored are the ones that are most significant to ensure reliable operation. In the case of some significant latent defects, the measurements of critical parameters may be within an acceptable range, but over time fatigue damage or chemical reactions can degrade some parameter that is critical to a system's operation. The distribution of the product's initial manufactured strength, which is defined by its operational capability and mechanical robustness, may vary throughout the production. As with any prediction of the future events, it is difficult to predict which manufacturing parameters will have the greatest variability, as well as when they will occur and by how much.

Component failure does not necessarily lead to system failure. In digital systems a component with a latent defect may lead to degraded operation, or it may not affect the system operation at all. How much variation in a component a system design can tolerate is unique to that system's design. In some applications of the same component, a circuit or system may tolerate wider excursions of the component while in other applications the variation or outright failure may not be observable in the effect on the circuit or system.

Of course, the trade-off is that we must consider the competitive market for which the product is being developed. Companies with market competition must build the unit at the lowest possible cost and get to market with innovative designs first. In some cases, special materials may be necessary for certain harsh environmental applications such as the electronics used in oil and gas well drilling operations.

The environmental stresses for electronic systems used thousands of feet below ground are some of the most extreme for any electronics equipment. Measurement while drilling (MWD) electronics components used in downhole applications must not only operate at temperatures exceeding 200°C, but also be extremely reliable; equipment failure leads to rig downtime, which is often extremely costly. At these required high temperatures, standard solders and PWBs cannot be used, although many of the active devices used are rated for 80°C by the manufacturer. They are being uprated and qualified for these applications because the needed components are not produced for this environment and because they can and have been reliably uprated for these applications. It is another demonstration of how significant the reliability entitlement of the silicon die is in ICs in extremely high temperatures.

Since not many companies are actually testing to raw thermal empirical stress operational limits, few know what the real operational margins can be in current electronic systems. The author has applied thermal HALT to many complex electronic systems and discovered that many can operate from −60°C to +130°C or greater margins, even using standard components and materials. The operational limits of electronics systems will likely change as new materials and processes for electronics are introduced, but testing to limits will be the quickest way to determine and benchmark stress limits and therefore the stress margins that are possible to achieve to ensure a robust product.

2.11 What Does the Term HALT Actually Mean?

The term HALT is commonly used to describe a type of environmental chamber that has the capability of rapid thermal cycles and simultaneous multi-axis pneumatic repetitive shock vibration. It is important to remember that HALT is not a specific type of test chamber, but instead it is a basic methodology of applying a stress or combinations of stresses to find operational and mechanical strength limits against using a variety of stressors.

HALT may be performed with a variety of thermal, voltage, vibration and other stresses that are relevant to the product's potential reliability failure mechanisms. It is not necessary that the stresses that are used in HALT exist in the end-use environment. If a stress used in HALT finds a relevant weakness or defect, it is valid, whether or not the product will experience the stress in the end-use environment. A solder crack is a good example of a defect that is stimulated by both thermal cycling and mechanical vibration and applying both stress stimuli simultaneously significantly accelerates the detection of the flawed solder joint, and therefore these are valid HALT stresses, even if the product is stationary or constant temperature.

There are stresses that do not require a HALT chamber. A HALT can be performed using increments of high and low voltage stress and high and low clock frequency stress to discover the operational limits. Thermal HALT testing can be performed in conventional thermal chambers and a classical electrodynamic shaker can be used for vibration HALT, but in general they do not have the rapid thermal cycling capability or high shock pulse peaks that induce fatigue damage at a higher rate as a HALT chamber with multi-axis pneumatic hammers does. A HALT chamber using liquid nitrogen for cooling and large electrical resistive banks for heating is the most capable and effective type of chamber to perform HALT, and even more so for HASS in manufacturing where testing throughput is critical to cost effectiveness.

There are significant benefits in using the HALT chambers in that they are capable of forcing rapid thermal transitions on an operational system under test. Higher thermal transitions produce higher thermal differentials and therefore thermal mechanical stresses across circuit board components and material bonds. Thermal cycling is very beneficial when used to perform HASS. In HASS the goal is to rapidly induce the highest combinations of stresses to precipitate latent defects that may occur during manufacturing to a detectable state. The more stresses that can be combined in HASS the more comprehensive the stimulation of latent defects with higher screening strength for faster detection.

The goal of HALT is to find out how close the strength of the design is to the fundamental limit of technology. The FLT is the point at which the design capability cannot be increased with standard materials. Some technologies, such as the operation of an LCD display, have a relatively low operational limit due to the physical properties of the liquid crystal. At temperatures greater than about 70°C, a standard LCD display will become dark with no contrast, but it will recover and operate when the temperature again falls below that temperature. If an LCD is used for monitoring system operation during HALT, it may be necessary to find alternative indicators of system operation or extend the monitor outside the test chamber. Sometimes designs have significant thermal operating margins without modifications, and the HALT limits can then be used to design shorter and more effective combined stress HASS tests to protect against manufacturing excursions that result in latent defects.

In most applications of electronic systems, technological obsolescence comes well before components or systems wear out, especially in the consumer markets such as cell phones and personal computers. We will never empirically confirm the total intrinsic life entitlement of most electronic systems since very few systems are likely to be operational long enough to determine the failure rates when intrinsic 'wear-out' failures occur. Again, it is important to emphasize that we

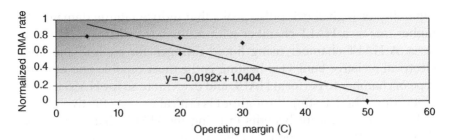

Figure 2.18 Cisco normalized return rate versus thermal operating margin. Source: Kyser, 2003. Reproduced with permission of NP Communications LLC/EE-Evaluation Engineering

are referring more to the life of solid state electronics and less to mechanical systems where fatigue and material consumption are the main contributors to wear-out failures.

Reliability tests and field data are rarely published but there is one published study with data showing a correlation between empirical stress operational margin beyond specifications and field returns. In 2002, Ed Kyser and Nahum Meadowsong from Cisco Systems gave a presentation entitled 'Economic Justification of HALT Tests: The relationship between operating margin, test costs, and the cost of field returns' at the IEEE/CPMT 2002 Workshop on Accelerated Stress Testing. In the presentation they showed a scatter diagram comparison of thermal stress operational margin versus the normalized warranty return rate on different line router circuit boards with a best fit trend line of data, as shown in Figure 2.18. Normalized RMA declines as thermal margin increases.

The graph shows the correlation between the thermal margin and the RMA (return material authorization), i.e. the warranty return rate. A best fitting curve with this scatter diagram shows a probabilistic mathematical relationship between thermal operational margin and warranty returns. It indicates that the lower the operational margin, the higher the probability of its return. Cisco also compared the relationship between the number of parts (on a larger range of products) and the return rate. The graph of that data is shown in Figure 2.19. The relationship between thermal margins versus return rates is ten times stronger than the relationship between board part counts versus return rates. If all conditions of development, manufacturing and end-use environmental conditions are to be the same for future systems it

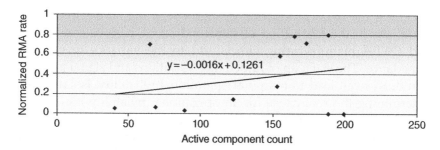

Figure 2.19 Cisco normalized RMA rate versus active components count. Source: Kyser, 2003. Reproduced with permission of NP Communications LLC/EE-Evaluation Engineering

would be possible to estimate the probabilistic decrease in RMA rates per increase in thermal margin.

A better use of the relationship between RMA rates and thermal operational margin may be to determine the ROI of a change in the initial design to increase the margins to reach the highest level of operational margin, which in this case is 50°C above product design operational specifications. If we know the costs of RMA for these systems we can make a business case for the ROI estimates for the cost of design or component changes that might be implanted to increase a thermal operating margin. The RMA savings cost in terms of dollars per degree centigrade can be used to determine the ROI of a proposed change to increase margins.

This makes sense in terms of the stress–strength relationship. As mentioned earlier, no matter how long a chain is, it is only as strong as the weakest link in that chain. No matter how many active parts there are on the PWBA, the design's tolerance to variation in manufacturing and end-use stress is dependent on the part least tolerant to thermal variation, which translates to the component with the lowest stress operational margin.

For soft failures (i.e. not catastrophic hardware failures) that can be power cycled or rebooted to return to normal operation, the relationship between thermal limits and field operational reliability is less obvious. Because most electronics companies do not discover and therefore do not compare empirical thermal limits with rates of warranty returns, the relationship shown in Figure 2.18. may never be observed or used to increase operational reliability.

During the mass production of high speed digital electronics, the variations in the fabrication of semiconductors, passive components and PWBA manufacturing effects add to the variance in electrical impedance. The stack up of impedance variations and signal propagation delays can lead to marginal operational reliability. Isolating and determining the root cause for an operational reliability issue that is marginal or intermittent is challenging. These failures are not reproducible on the bench when tested and therefore are considered CND (can not duplicate) or NFF (no fault found) returns. They may be sent back to repair depots to be used for a replacement part in a warranty repair depot, potentially resulting in passing on the marginal problem to another customer.

Many times the marginal operational failures observed in the field can be reproduced when the system is cooled or heated to near the operational limit. Heating and cooling the system skews the voltage, impedance and propagation of signals, which also occurs in the variations in electrical parametrics from a stack-up of process variations during mass manufacturing. If companies do not apply thermal stress to empirical limits, they will never discover and be able to utilize this benefit to find difficult to reproduce signal integrity issues. This aspect of thermal stress and the benefit of the discovery of software failures will be covered in more detail in Chapter 7.

Faster testing using higher stresses, which therefore costs less, also results in quicker discovery of risks to reliability. Reducing test time and costs is becoming more critical in today's accelerating pace of new electronics product development. Most conventional reliability testing is done to a pre-established stress above spec or using a 'worst case' field stress, which may take many weeks to several months. Both result in minimal reliability data. Finding an electronic system's strength by HALT methods is relatively quick. In most cases, to find thermal and vibration limits using HALT takes a week or less and uses fewer samples than does end-use environment simulation testing.

When the product being evaluated in HALT meets the goal of reaching the capability of the FLT, there is still useful variable data from operational and sometimes destruction limits. Empirical stress limits will vary between samples. If the operational limits between samples of the same precisely configured system are close together, say within $\pm\,10°C$ for thermal and $\pm\,5$ Grms for vibration, when tested under the same stress levels and durations, then it is likely that the design and manufacturing are consistent and capable. Significant variations of strength limits between samples of the same system can be an indicator of some underlying inconsistency in the manufacturing process. If the variations are large enough, a certain percentage will have a high probability of unstable operation or failure because the end-use stress conditions exceed the variation in the product's strength.

Empirically discovered stress limits in an electronics system design are very relevant to potential field reliability and especially when comparing thermal stress limits to operational reliability in digital systems. Not only can the stress limits discovered in HALT be used for making a business case for costs of increasing thermal or mechanical margins, but also the data can be used for comparing the consistency of strength between samples of the same product.

As with any major paradigm shift, in the move from using the dimension of time to the dimension of stress as a metric for reliability estimations, there will be many details and challenges yet to be determined on how best to apply it and use the data derived from it. Yet, from a physics and engineering standpoint, a new reference of stress levels as a metric for reliability projections and comparisons has much greater potential for relevance and correlation to field reliability than the previous FPM that uses broad assumptions on the causes of field operational and hardware unreliability in current and future electronic systems.

When we begin using the empirical strength of materials and the understanding of the physics of stress limits and the combinations of stress limits as a new reference for reliability assessments, we will develop better test regimes and find better reliability performance discriminators, which will result in improving real field reliability at the lowest costs.

Bibliography

[1] Black, J. R. *Mass transportation of aluminum by momentum exchange with conducting electron*. 1967. Proceedings of the Sixth Annual Reliability Physics Symposium. pp. 148–159.

[2] Orio, R.L. de. Dissertation: *Electromigration Modeling and Simulation*. [Online] June 2010. [Cited: 8 Aug 2014.] http://www.iue.tuwien.ac.at/phd/orio/node16.html.

[3] Rosenberg, C.-K. Hu. *Scaling effect ion electromigration in on-chip Cu wiring.* and R. 1999. IEEE International Interconnect Technology Conference. pp. 267–269.

[4] Srinivasan, J., Adve, S.V., Bose, P. and Rivers, J.A. *The case for lifetime reliability-aware microprocessor.* 2004. IEEE Proceedings of 31st Annual International Symposium on Computer Architecture.

[5] Karri, A., Dasgupta, R., *Electromigration reliability enhancement via bus activity distribution.* Las Vegas, Nevada: s.n., 1996. 33rd Design Automation Conference.

[6] Suzuki, E. Takeda, N. *An empirical model for device degradation due to hot-carrier injection for device analysis.* IEEE Electron Device Letters. 1983, Vols. EDL-4, pp. 111–113.

[7] JEDEC. *Failure Mechanisms and Models for Semiconductor Devices.* s.l.: JEDEC Solid State Technology Association, 2003.

[8] Chakravarthi, S., Krishnan, A.T., Reddy, V., Machala C.C. and Krishnan, S. s.l.: IEEE, *A comprehensive framework for predictive modeling of negative bias temperature instability.* 2004. 42nd International Reliability Physics Symposium. pp. 273–282.

[9] Wywas, E.J. and Bernstein, Joseph B. *Quantitatively Analyzing the Performance of Integrated Circuits and Their Reliability.* IEEE Instrument & Measurement Magazine. 2011, February.

[10] Gray, K., Pecht, M. *Long-Thermal Overstressing of Computers.* IEEE Design and Test of Computers. November/December, 2011, Vol. 29, 6, pp. 58–64.

[11] Boeing 787 Dreamliner battery problems. *Wikipedia.* [Online] 30 Jan 2014. [Cited: 31 Jan 2014.] http://en.wikipedia.org/wiki/Boeing_787_Dreamliner_battery_problems#cite_note-55.

[12] Apple, Inc. Environment Requirements for the iPhone 5. *Apple iPhone 5 – View all the technical specifications.* [Online] 2013. [Cited: 2 May 2013.] http://www.apple.com/iphone/specs.html.

[13] Vichare, N., Rodgers, P., Eveloy, V., Pecht, M., Environment and Usage Monitoring of Electronic Products for Health Assessment and Product Design. 2, 2007, *Quality Technology and Quantitative Management,* Vol. 4, pp. 235–250.

[14] M. Pect and F. R. Nash, "Predicting the Reliability of Electronic Equipment," *Proceedings of the IEEE,* 1994.

[15] DOD Guide for Achieving Reliability, Availability, and Maintainability, 2005.

[16] Denson, W. The History of Reliability Prediction, *IEEE Transactions on Reliability,* vol. 47, no. 3, pp. 321–328, 1998.

[17] Decker, G. *Policy on Incorporating a Performance Based Approach to Reliability in RFPs.* 15 Feb 1995.

[18] Harms, J. *Revision of MIL-HDBK-217, Reliability Prediction of Electronic Equipment.*, San Jose, CA: s.n., 2010. Reliability and Maintainability Symposium (RAMS). pp. 1–3.

[19] Pecht, M., Dasgupta, A. and Leonard, C.T. The Reliability Physics Approach to Failure Prediction Modeling.4, 1990, *Quality and Reliability Engineering International*, Vol. 6, pp. 267–273.

[20] McCluskey, P., "Reliability of Power Electronics Under Thermal Loading," *Proceedings 2012 7th International Conference on Integrated Power Electronics Systems (CIPS)*, Nuremberg, 2012.

[21] Pecht, M., Das, D., Ramakrishnan, A., The IEEE Standards on Reliability Program and Reliability Prediction Methods for Electronic Equipment, *Microelectronics Reliability*, vol.42, no.9–11, pp.1259–1266, September-November 2002.

[22] Lall, P., Harsha, M., Kumar, K., Goebel, K., Jones, J. and Suhling, J., *Interrogation of Accrued Damage and Remaining Life in Field-Deployed Electronics Subjected to Multiple Thermal Environments of Thermal Aging and Thermal Cycling*, Electronic Components and Technology Conference, 2011.

[23] Pecht, M. Why the traditional reliability prediction models do not work - is there an alternative?, *ERI Reliability Newsletter*, vol. 4, 2001.

[24] Cushing, M., Mortin, D., Stadterman, T. and Malhotra, A., Comparison of Electronics-Reliability Assessment Approaches, *IEEE Transactions of Reliability*, vol. 42, no. 4, pp. 600–607, 1993.

[25] Jones, J, and Hayes, J., A Comparison of Electronic Reliability Prediction Models, *IEEE Transactions on Reliability*, vol. 48, no. 2, pp. 127–134, 1999.

[26] Brown L., *Comparing Reliability Predictions to Field Data for Plastic Parts in Military, Airborne Environment*, Reliability and Maintainability Symposium, 2003.

[27] Smith, C. and Womack, J. *Raytheon Assessment of PRISM as a Field Failure Prediction Tool*, Reliability and Maintainability Symposium, 2004.

[28] Jensen, F., Peterson, N. *Burn-in*. New York: Wiley and Sons, 1982.

3

Challenges to Advancing Electronics Reliability Engineering

3.1 Disclosure of Real Failure Data is Rare

The study and advancement of reliability engineering of electronic systems has inherent limitations when trying to correlate theory to actual case histories and field data.

In the world of competitive electronic systems, data on the field reliability of an electronic system is one of the most guarded secrets of electronics manufacturing companies. The major limitation is that the real root causes of failures and rates of field failures are rarely, if ever, published if companies are not forced to by court order, which rarely occurs.

Advancements in any field of engineering improvements and innovation in technology are developed by building on previous knowledge using observation and analysis of empirical evidence. In the field of reliability engineering – and in particular electronic assemblies and systems – knowledge about field failures of electronics hardware and the true root causes is extremely limited. For the most

Next Generation HALT and HASS: Robust Design of Electronics and Systems, First Edition.
Kirk A. Gray and John J. Paschkewitz.
© 2016 John Wiley & Sons, Ltd. Published 2016 by John Wiley & Sons, Ltd.

part, this is due to the sensitivity of failure data in a competitive market. If a product failure results in monetary loss or, in the worst case, an injury to a person, the criminal or liability costs for a company could be significant.

Without the ability to share data and teach what we know about the real causes of unreliability in the field, we cannot show the evidence for a new reliability methodology or reliability test approach that is making a real difference in field and warranty costs.

With this lack of distribution of knowledge about the majority of real root causes of electronic system failures, it is easy to understand the continued belief in the potential ability to model and predict the future of electronic systems lifetimes and rates of failure. Many companies manufacturing electronics still specify reliability requirements in terms of averages such as MTBF, which have little or no traceability to the physics that results in failures in electronics assemblies.

There are many mechanical and electromechanical devices with moving parts in many electronic systems. Switches, motors, mechanical relays and connectors are mechanical devices that have intrinsic fatigue damage or material consumed over time which leads to wear-out failures. Yet, material wear or fatigue damage progress can be measured as they decay, and models can be developed for these mechanisms to ensure that there is enough material strength or needed reservoir of material in the design to meet the intended use-life requirements. When hardware becomes available, the models used to estimate the rate of material consumption in mechanical interfaces can be measured to ensure that rates concur with the models. Since no materials are consumed in solid state electronic systems, degradation mechanisms are much more difficult to model. There is a new emphasis in reliability engineering of electronics to find and use better parametric discriminators for in-situ prognostics and health monitoring (PHM) of electronic systems to determine the remaining useful life (RUL)[1].

Often when there are new engineering developments or approaches to testing that provide new benefits case studies and empirical evidence of the benefits are published. There are several reasons why details of case histories of HALT successes or errors in the application of HALT (as well as any other actual empirical electronics reliability field data) are rarely published.

Some key reasons are:

- competitive advantages: companies do not disclose the most effective reliability practices
- potential corporate legal liability from field failures
- avoiding blame: the true cause can be difficult to determine because product failures can be very expensive and damaging to a company's reputation
- engineers have little time to write and publish successes, and much less motivation to write about causes of failure.

When a company discovers a new product development process that leads to significantly faster times and lower costs to release a mature product to market, they are not likely to tell their competitors. Doing so would cause them to lose the competitive advantages of those new processes, so little data has been published or will be in the future on HALT and HASS reliability development successes.

Legal liability for product failures can be a large economic risk for electronics manufacturers, depending on the product and application. Failures of electronic systems might lead to loss of property or even injury or death. Disclosing the root cause of failures of electronic systems could lead to loss of a company's quality and reliability reputation. It may also provide evidence of a manufacturer's liability from design or manufacturing errors that would result in costly court judgements against them.

Because of the sensitivity of electronics failures, reliability engineers who may want to help the field of reliability engineering by publishing case histories of reliability issues are generally under various restrictions. If they are able to make the time to write a report for publication or public presentation showing a cause of field failures, they will face many challenges to persuade the legal department of their company to give permission to publish any evidence of errors in design or manufacture. Even if they are able to publish something on actual reliability, the paper will be so redacted and 'sanitized' for public disclosure so that the most significant and relevant data may not be published. Unless engineers are willing or able to publish real case histories, details of the root causes of the failures, and the best methods to prevent them, little can be expected in the advancement of the science of electronics reliability development and testing.

If reliability engineers investigate, analyze and understand the root causes of field failures in their own products, they would see that most verified failures come from assignable causes that can and should be prevented. If a company does not know what is causing unreliability in their products, how can they possibly improve reliability?

The underlying causes of failure in electronics can be complex and many times it is due to a sequence or combination of events that individually would not have resulted in failure. The verified failures returning from the field confirm that the time and degree of impact of each event that creates the latent defects leading to field failures cannot be modeled or predicted.

We can still make progress in the field of electronics reliability and understanding intrinsic modes of degradation over time, but we must validate the methodology and results from a material science and engineering basis. We must use our knowledge of physics and material science along with the lessons we have learned from real causes of field failures.

3.2 Electronics Materials and Manufacturing Evolution

Electronics materials and manufacturing methods are rapidly evolving. Electronic systems will continue to decrease in power, and while semiconductor device densities and system integration increase. The pace of change makes it extremely difficult to analyze and model intrinsic wear-out failure mechanisms from new materials and processes.

An example of a new failure mode in electronics assemblies occurred due to the Reduction of Hazardous Substances (RoHS) directive which was adopted by the European Union. The RoHS directive forced electronics manufacturers to change from using solders containing the element lead to lead-free solders that were mostly mixtures of tin, silver and copper. A failure mechanism observed and documented in the early 20th century using tin solder in the manufacture of vacuum tubes resulted in the growth of fine filaments out of the solder – referred to as 'tin whiskers.' The growth of tin whiskers was revealed to be a reliability problem after the change to lead-free solders. Tin whiskers are known to have caused many military and aerospace systems failures and are suspected to have caused unintended acceleration in Toyota automotive control systems, although it has not been acknowledged by manufacturers of the systems [2]. Adding lead to the solder prevented the growth of the tin filaments that had led to shorting between adjacent conductors. When lead-free solders were used because of RoHS, electronic system failures caused by the tin whiskers began to occur in the field. The problem has been mitigated, but it took many failures to discover.

It is important that there is traceability, causality and empirical validation of the modeling of physical failure mechanisms for electronics reliability models of wear-out mechanisms now and in the future. For instance, traditional electronics FPM (failure prediction methods) have used the Arrhenius equation and the broad assumption of 0.7 eV for the activation energy leading to the intrinsic failure of a silicon IC. Unfortunately, the use of the activation energy value of 0.7 eV for semiconductors continues today, even though there is little or no reference to a specific physical mechanism for this relationship.

When we find a weakness in an electronic system through stepped-stress methods, we should know enough about the materials to conclude that the weakness is due to a fundamental limit of technology (FLT), such as the melting of plastics, reflow of solder or limits of LCD operation at temperature or if the weakness is due to the in-circuit application of a particular component. After uncovering the causes, we can understand what physics drove the failure and know the element to change to increase the system's strength or capability.

Usually, it is only necessary to strengthen one or two of the weakest elements in the design to bring a product's strength up to the FLT. Sometimes software interactions with hardware lead to intermittent or marginal operation, and changing code may be the only change necessary to add significant thermal strength capability and margins.

Occasionally, the system is designed and built and reaches the stress FLT with no change needed, and this becomes a benchmark for subsequent designs. But if you are not testing to empirical stress limits, you will never discover the FLT nor use the high level of strength to find unreliable elements that can be introduced during the manufacturing phase.

The limitations on sharing real field and test lab reliability data are not likely to change anytime soon. However, we can change our approach to electronics reliability development based on a new orientation of finding the empirical strength of a system against the unknown life cycle stresses it may be subjected to.

We must acknowledge that the causes of most electronics failures in the first five to seven years of use in the vast majority of cases are not due to intrinsic or known wear out failure mechanisms in the devices, but are assignable to special causes that will support the allocation of reliability development activities of new electronic systems. Instead of continuing the unsupported belief that dominant causes of unreliability can be modeled and predicted, most work should be directed at discovering potential weaknesses in a new design and improving them when possible. Many electronics manufacturers have realized these facts and have shifted resources to proactive HALT and HASS methods, but they do not publicize their results because they do not want to educate their competitors on the most cost-effective reliability development techniques.

Bibliography

[1] *Prognostics and Health Management of Electronics.* Vichare, N., Pecht, M. 1, March 2006, IEEE Transactions on Components and Packaging Technologies, Vol. 29, pp. 222–229.

[2] Quality Control System Corp. How NHTSA and NASA Gamed the Toyota Data. Safety Research and Strategies, Inc. [Online] 21 July 2011. [Cited: 3 July 2013.] http://www.safetyresearch.net/2011/07/21/how-nhtsa-and-nasa-gamed-the-toyota-data/.

4

A New Deterministic Reliability Development Paradigm

4.1 Introduction

Considering the limitations of statistical reliability predictions and metrics covered in the previous chapters, it is the objective of this book to propose a more effective method of achieving product reliability in electronic systems. In this section, a new paradigm for reliability practitioners will be covered. It focuses on determining empirical limits and design margins of new products with accelerated stress testing and an understanding of the physics of failure mechanisms in the design using the HALT methodology that was introduced in Chapter 1. It is integrated into the product and production process development process to help guide its implementation. This enables us to apply effective corrective actions and ensure robust products capable of performing reliably when exposed to the variable stresses in the operating environment and the inevitable variability in supplier materials and product strength.

Next Generation HALT and HASS: Robust Design of Electronics and Systems, First Edition.
Kirk A. Gray and John J. Paschkewitz.
© 2016 John Wiley & Sons, Ltd. Published 2016 by John Wiley & Sons, Ltd.

Instead of drawing upon a general database that is likely not to be representative of the new product or operation environment, our proposed method draws upon the knowledge baseline of previous similar products and field failure data and then focuses on the changes needed in the new product being developed. This allows analysis and testing to be targeted on the unknown and changed features needed in the new product. Few products are truly entirely new and innovative. Most products are evolved from previous designs and use technologies proven in past applications. So, the focus is on the changes needed for the new derivative product or application. Changes that prompt designs to evolve include new regulatory requirements and standards, availability of new materials or technologies that may be applicable to our needs, and new market expectations that require improved and updated solutions.

This chapter overviews the remainder of this book and integrates the tools and their application into an overall process for product and process development aimed at creating robust and reliable products within the constraints of the modern product development environment. This includes the need to launch new products in a shorter time period and at reduced cost. These challenges preclude the use of some of the traditional approaches and they require us to use focused methods that create reliable and capable designs in minimum time. It also requires an understanding of the limits and physical mechanisms causing failures in design choices. The tools and processes presented in this book are aimed at accomplishing this learning and verification of the design and production process as efficiently and effectively as possible.

To help introduce this proposed method, a product and process development flow using the recommended tools is shown in Figure 4.1. This helps the reader visualize the process and see how the tools are integrated and work together to ensure robust and reliable design of products and production processes. Each of the following chapters focus on a tool and highlights where and how to apply it during product development. The process begins with understanding customer needs and the application environment. Risks are assessed based on considering baseline knowledge of previous and similar products, field failure data and anticipated design changes and unknowns. Design for reliability and physics of failure are key parts of design, analysis and test phases to identify and correct weaknesses. After a robust design, simulation

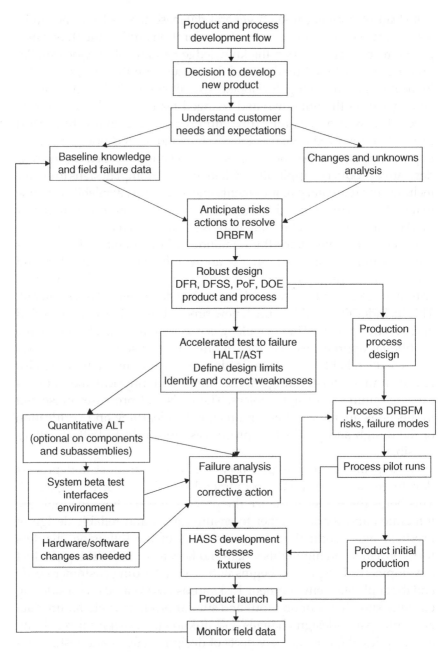

Figure 4.1   New product and process development flow

and analysis of the new product, prototypes are constructed and evaluated in HALT to find design limits and correct weaknesses found. The production process design begins in parallel with the product design, and integrated teams help to ensure that producible design choices are made to reduce product variability. Quantitative accelerated life test can then be used to establish life versus stress trade-off curves, assess degradation and estimate reliability for satisfying customer expectations. HASS is used to screen production products to detect latent defects and variability in production to ensure that only robust and defect-free products reach the customer.

All of these tools together form a focused and efficient approach to new product development that meets customer expectations in less time while preventing costly problems in field application. The following chapters explain the tools, processes and sequence used to accomplish this.

4.2 Understanding Customer Needs and Expectations

Reliability in product development begins with understanding customer expectations. What is the operating environment that the customer will operate the product within? What range of stresses may be applied? What is the nominal use stress and severe use stress applied to the product? Next, the customer may indicate an expected level of reliability for the product. This may be based on preventive maintenance cycles or failure free run times and warranty considerations. These levels of reliability may be expressed explicitly as a percentage reliability over a specified period of operating time under specified operating conditions, or as a mean life or a minimum life under certain operating conditions. In other cases, the customer may have implicit reliability needs such as a failure free operating period, a specified warranty period, life-cycle cost, maintenance interval or reliability as good as or better than the legacy product. Lastly, there are customers who are unsure of their reliability needs, and the supplier must uncover those needs. It is important here to include the severe user as well as the nominal user in planning for reliability, so that the range of operating conditions can be understood. This discussion may also reveal unknowns about operating conditions and the need to provide more design margin to cover these unknowns. For industrial and defense customers, the operating conditions can often be defined more clearly during this dialogue, including sensor measurement data, infrared thermal analysis and field operation data. For consumer products, the operating environment is harder to define and anticipate, and it may require considering worse case and severe user scenarios. The knowledge and data of life cycle stress distributions are typically not available in consumer electronics, and future applications are not defined during the design phase. So this requires anticipating worst case scenarios during the risk assessment process.

A good way to explore customer reliability needs is through the voice of the customer meetings to understand the environment that the product operates in and the implications of product failure. The results can be documented in a Lean quality function deployment (QFD) matrix. This matrix is condensed from the traditional 'house of quality' QFD tool. An example of the format of the Lean QFD is shown in Figure 4.2 [1].

The Lean QFD helps identify targets and features to meet the customer's needs and provide designers with information they can design to. It is always best to involve customer representatives or user groups in this process. If that is not possible, marketing representatives who have been in contact with customers can provide useful insights into customer experience with current products and features they are

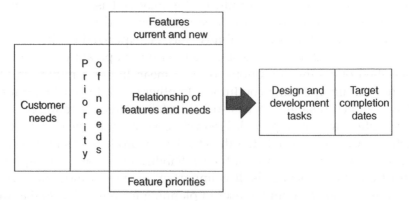

Figure 4.2 Lean quality function deployment chart to translate needs to design features [1]. Source: Adapted from Bechtold, 2011

seeking in new products. Needs can be prioritized using tools such as paired comparison or analytical hierarchy matrix to rank the needs in order of priority. The relationship between needs and features and the targets for each are essential for ensuring that the product meets the needs. The translation from needs to targets provides the information needed for engineers to design the product and accomplish learning and exploration during product development.

In evaluating the customer needs and determining design targets, conflicting needs may be revealed, and these require trade-off decisions to balance conflicting priorities as well as to arrive at solutions for features that adversely affect each other. Consider the following when prioritizing work on particular features:

1. What is absolutely essential?
2. Areas of potential compromise/trade-off
3. True needs for targets and features
4. Possible alternatives and a better solution to the customer's problem

From a reliability perspective, it is essential that the customer's reliability expectations and needs are addressed in this process. The operating environment, nominal and extreme user stresses applied to the product, maintenance intervals and impact of failures on operating performance all need to be understood. This is an opportunity to learn from previous field failures and understand the failure mechanisms that the product may be subject to. However, it is not always possible to anticipate all operating conditions because the product may be in a system that is later modified or used in an application not anticipated during the design and development phase. Understanding interfaces of components and subsystems is essential as these can be the source of many reliability problems. As a result, the process will benefit from accelerated testing to failure and design of experiments to evaluate materials and components before selection. Because all of the unknown stresses or load conditions cannot be anticipated, building a product with sufficient design margin to withstand later changes in operating conditions is essential. HALT and AST (accelerated stress test) can ensure that there is sufficient margin and capacity to handle stresses that were not envisioned during design.

4.3 Anticipating Risks and Potential Failure Modes

With the customer reliability needs understood, the next step is anticipating risks and potential failure modes. This starts with a high level top down risk assessment. Tools to accomplish this include Lean or top-down failure modes and effects analysis (FMEA) or design review based on failure modes (DRBFM). These methods focus on applying the knowledge base from previous product design and field failure data and the expertise of subject matter experts to anticipate potential problems starting at the concept level and updating the analysis as the design evolves. Unlike traditional FMEA, these methods are simplified and focused on preventing problems with the new product as well as focusing analysis and test work to understand failure mechanisms and product weaknesses. This analysis starts with considering changes from the previous product baseline. These changes are more likely to increase risk and precipitate new failure modes. New design features not previously used may also increase risk with new materials or the need to use new processes. So, the initial focus on changes and unknowns helps identify potential risks and failure modes that must be addressed with design and testing actions by the product development team.

Diagramming tools such as the parameter diagram, functional block diagram, boundary diagram and process flow diagram are useful in highlighting these changes and important interfaces that may be affected. In the parameter diagram in Figure 4.3, there are inputs that are controlled and selected to operate the system. The effects of input errors or incorrect control signals need to be considered. The noise factors may be an even bigger concern. Although information on the operating environment and conditions are identified, unknown supplier changes and unforeseen applications can significantly impact the performance of the product and its response or output. This is why the parameter diagram is helpful in prompting thoughts on potential failure modes of the new product.

The boundary diagram (Figure 4.4) helps identify interfaces between subsystems and components that may be potential contributors to failures in the product. These interfaces are often overlooked, as designers focus on a circuit board or assembly they are designing.

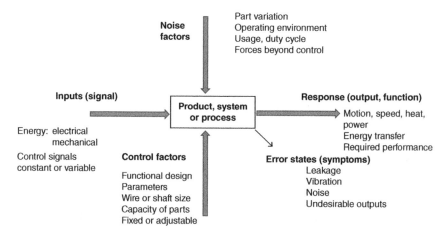

Figure 4.3 Parameter diagram elements [2,5]. Source: Adapted from Gokta and Ramamurthy, 2008 and Carlson, 2012

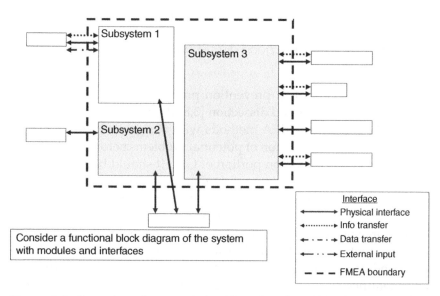

Figure 4.4 Boundary diagram derived from functional block diagram [2,5]

So, the diagramming tools help focus the risk assessment and help the participants to consider the less obvious aspects of the design that can sometimes be the root cause of later field failures.

DRBFM is a creative approach to FMEA developed by Toyota to get FMEA focused back on reliability problem prevention. It is part of the

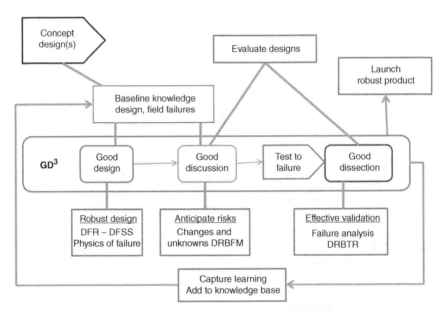

Figure 4.5   Good design, good discussion, good dissection integration [5,8,10]

Mizenboushi or problem prevention process, GD³, for good design, good discussion and good dissection [3,8] (Figure 4.5).

DRBFM combines FMEA methods with design review to focus on identification and correction of potential problems before they emerge. This is the good discussion portion of GD³. It should be started at the concept phase to assess relative risks of each of the concepts being considered. Then it is updated by the team as learning and design work progress. The method can be applied at the convergence/integration events or gate reviews used at each decision point in the product development process. DRBFM provides a review of the product concepts and design features by subject matter experts, as designs evolve, and identifies concerns that could become problems or failure modes as well as corrective actions to mitigate the risk involved (Figure 4.6). The corrective actions are tracked to closure. A significant difference from traditional FMEA is that scoring is not used to prioritize risk. Only the effect on the customer is considered, and it is rated high, medium or low, based on safety or regulatory compliance, failure to meet customer expectations or annoyance to the customer.

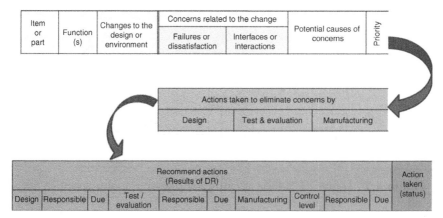

Figure 4.6 Basic design review based on failure modes (DRBFM) format [5,8,10]

All issues must be addressed, but higher impact items such as those affecting safety or product liability are addressed first. It is important to emphasize that risk assessment with DRBFM is a process and mindset where concerns and issues discovered in learning are captured in the DRBFM form as they are discovered by the project team and then reviewed prior to decisions. The DRBFM is applied in an integrated product and process design. It can be applied at the component, assembly, subsystem, system or production process level to anticipate and correct risks and problems identified. The format is only a means to capture the discussion and action items for follow-up. Completing the form is not the objective. It only documents the discussions and actions for use by the team to improve the product.

This risk assessment helps ensure that analysis and test work are focused on the areas of most concern for potential failure and performance issues. It is built on the knowledge base of previous product data and field failure information.

As work progresses from concept to more detailed design and comparison of design choices, the DRBFM approach can be applied at various levels of system development. It can be done at system level, part or component level, systems interfaces, and at the production process level (Figure 4.7).

Design or product FMEAs or DRBFMs and process FMEAs or DRBFMs are interfaced to ensure an integrated approach. The results of

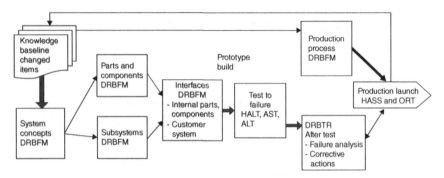

Figure 4.7   Levels of DRBFM in complex system development [3,5,8]

the DRBFM should contribute to test planning with corrective actions that identify need for testing to resolve the risks identified in the DRBFM. This can help with HALT and other AST planning to identify stresses to apply and items to be tested to determine design limits and weaknesses as well as to confirm potential failure modes identified in the DRBFM.

In process FMEA or DRBFM, a process flow map is the starting point to identify new or changed process steps for focused analysis. Key characteristics of the product that must be controlled in production can also be identified using the previous process DFMEA or DRBFM results and customer inputs on critical needs such as the interface of a supplier part or subsystem with an OEM customer system.

Similar to product designs, production processes are often built on standard methods or technologies. Many of these standard processes have been analyzed with a process FMEA, failure modes have been identified and corrected and controls of the process have been refined. With the need to produce a new product, these standard processes are the baseline and starting point to build the new product. So, once the new process steps or modifications to existing standard steps are identified, a process DRBFM focused on the new and changed steps can be helpful in understanding how the changes can cause the process to fail to produce the desired result or cause damage that weakens the product.

Figure 4.8 shows a typical process flow diagram for electronic circuit board manufacturing. This aims to identify which standard steps can be used for the new product and where changes are needed. Then the team can drill down into the new or changed steps to anticipate problems

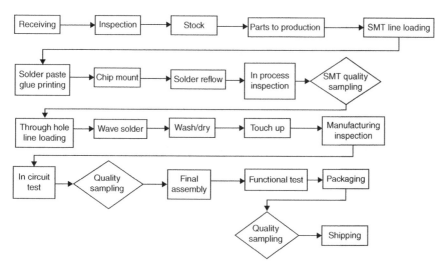

Figure 4.8 Typical process flow diagram

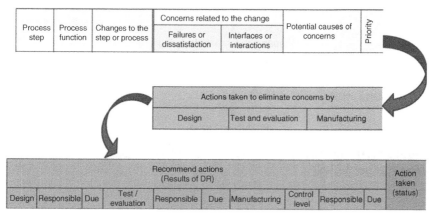

Figure 4.9 Typical process DRBFM format [5,8,10]

based on previous process issues indicated by statistical process control and cause analysis data.

Figure 4.9 shows the format for a process DRBFM. Just as in design, the process DRBFM targets identifying and preventing problems as new or modified production equipment and procedures are developed and refined. The results of this analysis are again updated as development proceeds and are reviewed at decision points in the process. The results can be used in selecting measurements to be made on the

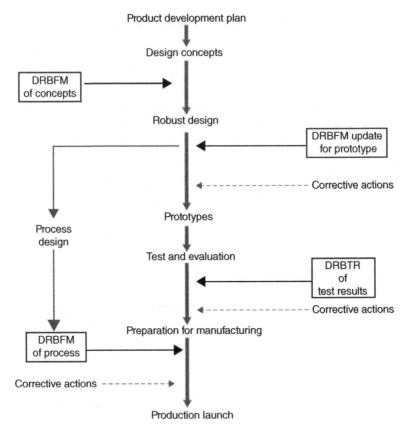

Figure 4.10 Timing and application of DRBFM in product and process development [3,8]

product during HASS or HASA (highly accelerated stress audit) to confirm the output of the new production process.

Figure 4.10 shows the timing and integration of DRBFM in the overall product development process.

4.4 Robust Design for Reliability

The next element of reliability is robust design. Tools such as Design for Reliability (DFR) and Design for Six Sigma (DFSS) can be used to address the physics of failure mechanisms and variability to make the resulting design capable of withstanding varying operating environment conditions despite variation in the strength of the production product.

Robust design + Controlled processes = Reliable product

Robust design methods include the following:

1. Increase strength of the part
2. Understand operating environment stresses
3. Select more robust parts or materials
4. Increase design margin
5. Supplement deterministic design with probabilistic tools
6. Reduce part strength variability
7. Understand sources of part variation and deterioration
8. Controlled production process (SPC – statistical process control)
9. Protect vulnerable components
10. Decrease effects of environment
11. Reduce unnecessary complexity of design
12. Function analysis
13. Value engineering techniques
14. Design for manufacturing and assembly
15. Design for maintainability/serviceability

Two key aspects of robust design are understanding the physics of failure and applying probabilistic design methods. The physics of failure involves identifying the physical mechanisms of failure of components in the system or subsystem and the stresses that precipitate these mechanisms. ASTs, like HALT, help to reveal these failure mechanisms and the stresses at which they occur.

The most common mechanisms causing failure include:

- corrosion or contamination
- wear at interface of moving parts
- mechanical failure (fatigue, vibration resonance, etc.)
- overstress (mechanical or electrical transient loads).

A list of primary failure mechanisms is shown in Figure 4.11.

Tools to evaluate these mechanisms in a design include finite element analysis (FEA), dynamic simulation of transients, fatigue analysis (cumulative damage), thermal analysis, and accelerated testing to failure.

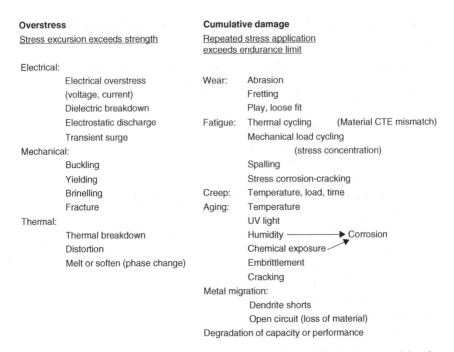

Figure 4.11 Summary of typical failure mechanisms to consider in design

FEA is used to evaluate mechanical and thermal stresses and response of components and assemblies to these stresses. This can help refine design features and materials selection to better withstand a range of applied stresses. FEA can also be used to assess the response of the unit to dynamic and transient stresses to better understand the response of the product and identify failure regions. Computational fluid dynamics (CFD) can be used to evaluate fluid flow and heat transfer effects on the product. Thermal analysis can apply infrared imaging to identify high temperature locations in the product and assist with component placement and cooling aspects of the design. All of these tools provide analysis and simulation to assist in the development of the design. After the concept design is created, dimensions and stress loadings can be modeled and applied to the designed component or assembly. Loads are simulated in the model and the response of the unit structure to mechanical, thermal and fatigue loads can be determined and displayed showing the location of the highest stress levels and their intensity.

Response to transients can be modeled as well. This simulation can reveal potential weaknesses in the design and reduce the number of prototype build and test cycles needed to develop the design. Design iterations can be quickly updated and rerun in the simulation. When the design has been refined sufficiently to produce acceptable results with the model and simulation, prototypes can be built and tested. Results of the testing can be used to update and validate the models used in the simulation.

Probabilistic design methods consider the variability of product strength and applied stresses and use this information to provide sufficient design margin for preventing failures.

These methods include:

1. understanding the physics of failure and stresses that precipitate failure
2. using of life-stress relationships and accelerated test to failure
3. considering variability of applied stresses and variability of product strength
4. eliminating stress–strength interference
5. allowing for degradation of part strength with repeated application of stresses.

The alternatives to reducing the stress–strength distribution interference are shown in Figure 4.12. These can include increasing the strength margin of the product, reducing product variability due to

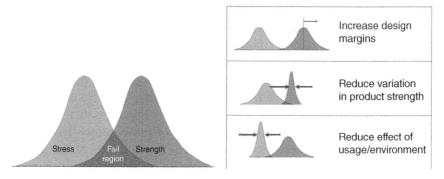

Figure 4.12   Reducing stress–strength interference [6]. Source: Vassiliou, 2008. Reproduced with permission of Reliasoft

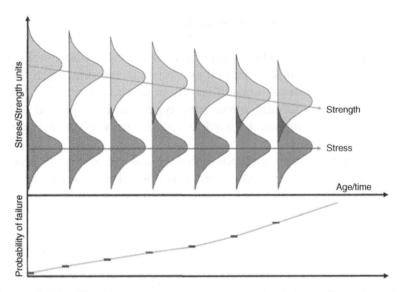

Figure 4.13 Effect of aging or wear on product strength and probability of failure [6]. Source: Vassiliou, 2008. Reproduced with permission of Reliasoft

process variation, or reducing the effect of stresses by protecting the product from environmental conditions.

The effect of repeated loading or stressing over time is shown in Figure 4.13. This represents cumulative damage and leads to wear-out failure mechanisms as strength degrades with repeated loading.

An illustration of probabilistic design is shown in Figure 4.14. Two design parameters result in intersecting bounds for the design. A deterministic solution is at the intersection of these two bounding factors. The probabilistic solution adds consideration of the variability of results and moves the entire population within the bounds to prevent failure. This produces a solution optimized for reliability with greater design margin to provide a more robust solution. Probabilistic design helps ensure that the range of variability in product strength is within the safe region.

The methods of reliability based design optimization (RBDO) to facilitate probabilistic design have been the subject of academic research [4,11]. Verifying these analyses in an actual design can be done with accelerated testing to help determine the margin and variation in actual product samples. These tools are more applicable to new materials and production methods that may be used in the design.

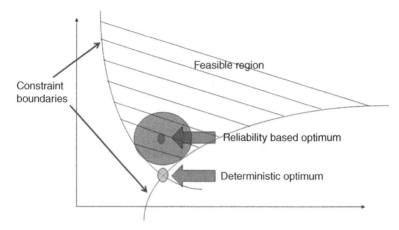

Figure 4.14   Illustration of a probabilistic solution using reliability based design optimization [4,11]. Source: Adapted from Agarwal, 2004 and Nguyen, 2010

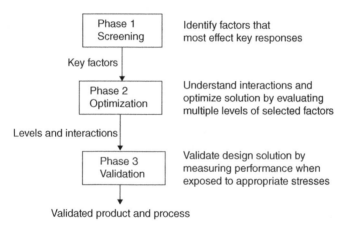

Figure 4.15   Phased DOE approach to optimize design choices [9]. Source: Adapted from Wachs, 2009

Design of experiments (DOE) is also an important element of robust design, and it is especially useful in product development for comparing components, materials and configurations of concept designs. A reliability focused DOE can be used to evaluate the effect of selected factors on the life of a component and aid in the selection of materials or purchased components. A multi-phased DOE approach, as shown in Figure 4.15, makes it possible to evaluate factors for the main effects in the first phase and then to understand interactions and optimize the solution in the second phase. The third phase of the DOE involves validating the selected solution by building and testing product exposed to the stresses anticipated. The use of multiple phases in a DOE reduces the overall number of samples needed as well as test time to complete.

4.5 Diagnostic and Prognostic Considerations and Features

Diagnostics and prognostics are also important design considerations. This part of the design effort looks at indicators of degradation and wear that can be monitored during operation of the product to anticipate problems. Once this is determined from testing, it can be used to develop the monitoring of key parameters during operation. This facilitates preventive maintenance planning to replace deteriorating parts before failure occurs. It also helps prevent unexpected shut down of the system or catastrophic failures.

Accelerated stress testing can use degradation to an acceptable limit as a definition of failure, even though the item is still able to function. This type of testing would be aimed at accelerating wear mechanisms identified in risk analysis and test planning. If key performance parameters are monitored and degrade to less than acceptable levels, the unit is considered to have failed. Analysis of degradation is illustrated in Figure 4.16. Such testing can help determine parameters to monitor with sensors for the diagnostic or prognostic subsystem to evaluate and communicate pending failure. This can facilitate preventive maintenance.

4.6 Knowledge Capture for Reuse

Results of the robust design, analysis, and DOE are captured for reference and reuse on subsequent designs. Methods to make this captured knowledge more accessible include the use of trade-off curves for materials, design configurations and components. Trade-off curves capture data points and provide a visual illustration of the relationship

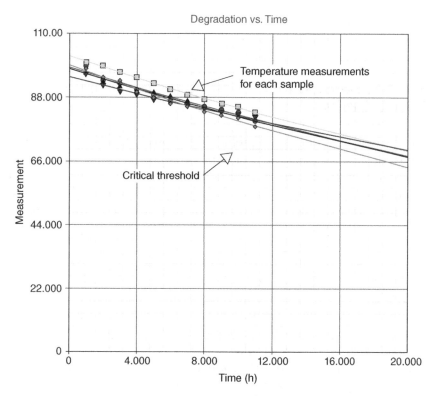

Figure 4.16 Illustration of degradation analysis. Source: Data from http://www.reliawiki.org/index.php/Degradation_Data_Analysis

between parameters. This knowledge is easily accessed for future work on similar products by other designers. From a reliability viewpoint, trade-off curves help to ensure that designs are well out of the failure region. The trade-off curves are based on test results from previous projects. The relationships between design parameters for various configurations of a current product reflect the learning on previous work and enable designers to understand the relationships of key parameters and make design choices that are more robust. Life versus stress plots from accelerated life testing can be used as trade-off curves to select allowable stress for a required life of a particular product configuration. At this stage, knowledge capture is focused on results of design, component and material evaluation having been completed during the design phase. As prototypes are built and tested, the results of testing and failure analysis are added to the knowledge capture. These are covered in more detail in section 4.11.

4.7 Accelerated Test to Failure to Find Empirical Design Limits

The next element of ensuring reliability in product development after robust design is testing the prototypes and production units to verify robustness of the design. Accelerated testing to failure is the primary element of this process. During the prototype phase, accelerated test to failure (HALT, step stress, specific stresses and failure modes, find material and component limits) is used to find design limits and weaknesses in the designed configuration. Accelerated testing is done in two phases. The first is accelerated stress test to find product limits when subjected to the selected stresses. The physics of failure analysis, FMEA/DRBFM and understanding of the operating environment including extreme user application are all considered in planning the testing. Prototypes tested in accelerated stress testing are subjected to stepped increase of the applied stress until failure occurs, which defines the upper limit of the product. It also helps in understanding failure modes and weaknesses of the product. Several samples are tested to help assess variability. This can be done using HALT, or the similar AST, for other stresses that may be applied. Another objective of this phase of testing is to find and correct product weaknesses discovered during the test. Examples of test profiles used for limit testing are shown in Figure 4.17 for an electric heater, and in Figure 4.18 for a sensor interface electronic module. Correcting weaknesses could include selecting more robust components or materials, changing a design configuration to resist damage (e.g. adding support to reduce vibration resonance) or protecting components from thermal or mechanical damage (e.g. adding insulation or protective covering).

The HALT methodology is covered in detail in Chapters 5 and 6 and form the core portion of this book. In this section, the role of HALT in robust product development is featured. The concept and basic methodology can be extended beyond electronics to a variety of other products using selected stresses appropriate for product and application. This extension of HALT is covered in Chapter 11.

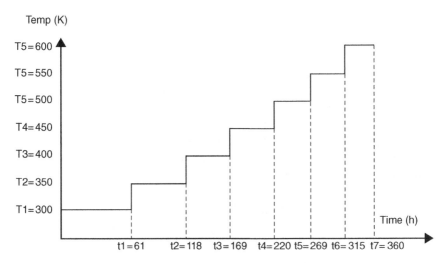

Figure 4.17 Step stress limit test profile

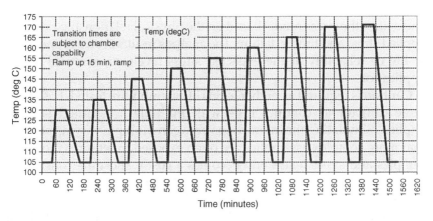

Figure 4.18 Step stress limit test profile for electronic modules with increasing temperature deltas

4.8 Design Confirmation Testing: Quantitative Accelerated Life Test

After weaknesses in the design are corrected, the design margin confirmation phase is the next level of testing used on selected components and subassemblies to confirm design margins and to estimate the reliability or the life of the unit when subjected to anticipated stresses. This phase uses the risk assessment used earlier as well as HALT/AST results in the previous phase to plan and conduct a quantitative accelerated life test using at least two or preferably three levels of the selected stress above the anticipated application use level but below the limits determined in HALT/AST during the prototype phase. HALT is a stimulation test to find design limits and weaknesses, but this next phase comes after HALT, and quantitative accelerated life test is a simulation to use accelerated test results to estimate life and reliability at anticipated application use level stresses. So, stress levels are selected between the limits found in HALT and the expected levels in actual use. This should be tested to failure at each of the accelerated test levels used in the test. This helps determine how and when the unit will fail when exposed to the expected stress levels. It involves testing to determine times to failure at the selected accelerated stress levels. The times to failure are analyzed using quantitative accelerated life test analysis tools. These methods fit the test data to a distribution and a life – the stress model based on the physical relationship between applied stress and time to failure. The analysis results in an extrapolation of life or reliability at the expected stress levels encountered in the application.

Caution is required as higher levels of assembly are tested. It is much more difficult to conduct tests with multiple accelerated stresses applied. Simulating the interactions between multiple varying stresses that more complex assemblies are often exposed to requires more test samples and stress levels and becomes a costly and time consuming test that may not be representative of field conditions. At higher levels of assembly, testing in the customer system and application may be needed to estimate life or reliability. See Chapter 9 for more details on quantitative accelerated life testing methods and considerations.

4.9 Limitations of Success Based Compliance Test

Compliance testing may also be required to confirm the ability of the product to meet industry standards or customer requirements for passing specified compliance tests. These tests can ensure that the product is able to withstand certain stress conditions, but these may not be representative of conditions actually encountered in operation. Success based compliance tests do little to confirm reliability of the product because no information on causes of failure or times to failure are determined during these tests. They only determine that the product can withstand a specified stress for a predetermined duration, but they provide no information on how and when a product will fail. Success based tests are sometimes called bogey tests and attempt to expose the product to one life at typical stress levels in the application. A larger number of samples and longer test time are typically needed, depending on the confidence level specified. With shorter product development schedules and limited test time and resources, it is more productive to conduct accelerated tests to failure to understand how and when the product will fail.

4.10 Production Validation Testing

When the product and process design is determined to be ready for production, additional testing is required to ensure that product generated with production tooling and processes is equivalent to the earlier prototype test results. Process variation and control must be measured to ensure latent defects and production issues do not degrade the performance of the product. Component and material supplier variation as well as internal manufacturing process variation can degrade the robustness and reliability of the product. One approach to this production validation stage is to use stress screening tests to detect weak products with latent defects before they are shipped to the customer. These stress screening methods include highly accelerated stress screening (HASS), environmental stress screening (ESS) and similar methods that apply stresses to production products at levels sufficient to detect weak units but still enable shipping of good units that have withstood the test stress. The tests are structured to not degrade life or performance of good units while detecting weak units.

Production validation is the last phase of robustness testing. The focus is on demonstrating that units built using production tooling, processes and suppliers perform similarly to prototype units tested during development. This testing also demonstrates that corrective action taken in design of the product or production processes is effective in preventing problems. Production units are subjected to stress screening to detect latent defects and variation in production and to screen out weak units so that they are not shipped to customers. Factors that were found to be discriminators during HALT can be used to monitor variation in capability and quality during manufacturing when applied in HASS. This prevents field problems and requires analysis and correction of the causes of the weaknesses and defects detected. HASS is a key tool in this process and is covered in detail in Chapter 7.

4.11 Failure Analysis and Design Review Based on Test Results

As each phase of robustness testing is completed, design review based on test results (DRBTR) is done to evaluate the effectiveness of the test and understand failure mechanisms. Like the earlier DRBFM to anticipate failure modes, the focus of the DRBTR is to evaluate test failures in detail and identify corrective actions needed to prevent problems. Corrective actions are assigned to responsible individuals and tracked to closure.

Robustness indicator figures, as shown in Figure 4.19, or stress boundary maps, shown in Figure 4.20, can be used to illustrate the margin in the design relative to expected stress levels in product applications. These can help facilitate decisions to proceed with development and production of the product or to perform additional corrective actions to the product or process to ensure the needed level of robustness.

Similar to the DRBFM method described earlier to anticipate risks and prevent problems, design review based on test results (DRBTR) is used to assess learning that took place during testing and identify corrective actions needed to correct deficiencies that remain in the product. Test results and observations are presented by test engineers with probable causes and comparison to previous testing of similar products. Reviewers are subject matter experts and make recommendation on closure of each finding [8]. Visual methods are used to indicate test and analysis results. One of these methods is the robustness indicator diagram shown in Figure 4.18 [7].

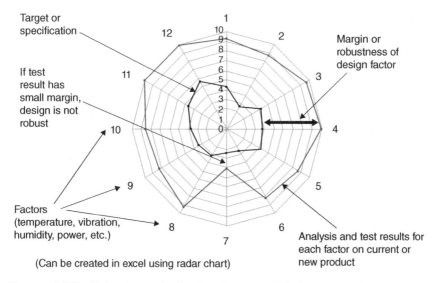

Target or specification

Margin or robustness of design factor

If test result has small margin, design is not robust

Factors (temperature, vibration, humidity, power, etc.)

Analysis and test results for each factor on current or new product

(Can be created in excel using radar chart)

Figure 4.19 Robustness indicator diagram [7]. Source: Adapted from SAE, 2009. Adapted with permission of SAE

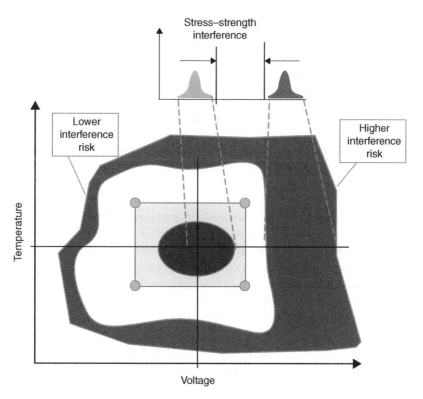

Stress–strength interference

Lower interference risk

Higher interference risk

Temperature

Voltage

Figure 4.20 Stress boundary map

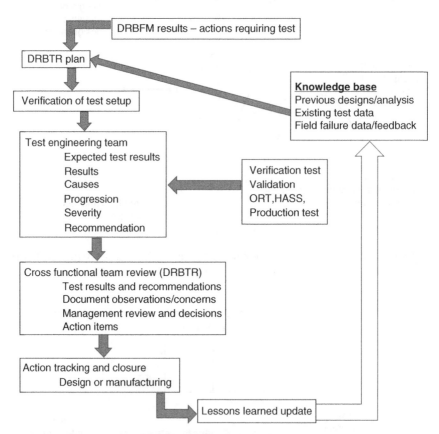

Figure 4.21 Design review based on test results (DRBTR) process flow [8]. Source: Adapted from Haughey, 2012. Adapted with permission of SAE

The DRBTR is the good dissection portion of the GD³ problem prevention process. The flow of the DRBTR process for review of test results in shown in Figure 4.21. Test results are summarized and compared to anticipated results as well as previous tests. Failed units are dissected and displayed for review by subject matter experts. Failure modes and mechanisms are discussed. Concerns and new questions may lead to new action items that require corrective action to the product or process to prevent the failure mechanisms that occurred. Figure 4.22 shows the format for capturing the DRBTR results and corrective actions after completion of a phase of testing.

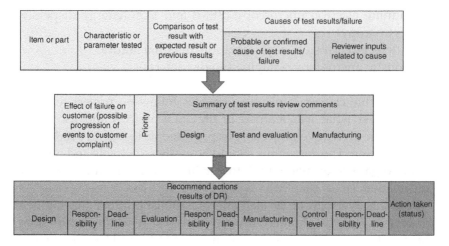

Figure 4.22 Design review based on test results (DRBTR) format [8]. Source: Adapted from Haughey, 2012. Reproduced with permission of SAE

Analyzing test failures to understand the mechanisms causing them is an essential part of evaluating test results. This is the good dissection part of GD³ using DRBTR. Failure analysis and reporting provide the basis for DRBTR and should be applied to testing at all phases of development.

Failure analysis is a progressive process that begins with documenting failure mode characteristics and observations, continues with non-destructive methods to examine failed parts, and finally dissection to expose failure damage and indicate failure mechanisms. Dissection risks damage to failure evidence, so gathering as much information with non-destructive methods first is essential.

- Basic information collection
  - recovery of failed samples
  - electrical test, microscopy, digital photography
  - infrared imaging during operation
- Non-destructive methods
  - X-ray (real time digital X-ray is particularly helpful)
- Disassembly/de-capsulation
  - tools or chemicals to remove layers

- Scanning electron microscopy (SEM) and EDS (energy dispersive spectroscopy)
  - defects, corrosion, contamination, material failure
- Acoustic microscopy/imaging (voids/defects)

Knowledge capture and reuse enables follow-on project teams to easily access what was learned in previous projects and use it as a baseline for derivative projects or improvements to extend product life or application. The data items should be easily searchable and retained in an organization-wide tool to facilitate locating and using the information.

With this overview of the elements of a robust product development process in place as an alternative to statistical reliability predictions, the next three chapters will focus on the HALT and HASS methodology and its application.

Bibliography

[1] Bechtold, L. (2011) Reliability Predictions to Support a Design for Reliability Program, IEEE 978-1-4244-8855-1/11.

[2] Gokta, S. and Ramamurthy, R. (2008), Power of P-Diagram in Robust Reliability Planning and Execution, Proceedings of The Applied Reliability Symposium, 17–20 June 2008, Reno Nevada.

[3] Shimizu, H., Imagawa, T. and Noguchi, H. (2003) Reliability Problem Prevention Method for Automotive Components, JSAE 20037158 / SAE 2003-01-2877, SAE International, Warrendale, Pennsylvania USA.

[4] Agarwal, H. (2004) Reliability Based Design Optimization: Formulas and Methodologies. Masters Thesis, University of Notre Dame.

[5] Carlson, C. (2012) Effective FMEAs, Achieving Safe, Reliable and Economical Products and Processes Using Failure Mode and Effects Analysis, John Wiley & Sons Inc., Hoboken, New Jersey

[6] Vassiliou, P. (2008) An Introduction to Design for Reliability, Proceedings of the Applied Reliability Symposium, Track 1, Tutorial 1, 17–20 June 2008, Reno, Nevada.

[7] SAE J1211 APR2009 (2009) Handbook for Robustness Validation of Automotive Electrical / Electronic Modules, SAE International, Warrendale, Pennsylvania USA.

[8] Haughey, B. (2012) Design Review Based on Failure Modes (DRBFM) and Design Review Based on Test Results (DRBTR) Process Guidebook, SAE International, Warrendale, Pennsylvania USA.

[9] Wachs, A. (2009) DOE for Reliability and Product Performance Optimization, Proceedings of the Applied Reliability Symposium, Track 2, Tutorial 2, 9–12, 2009 June San Diego, California.

[10] Allan, L. (2008) Change Point Analysis & DRBFM: A Winning Combination, Proceedings of the Applied Reliability Symposium, Track 1, Session 1, 17–20 June 2008, Reno, Nevada.

[11] Nguyen, T.H. (2010) System Reliability-Based Design and Multi-resolution Topology Based Optimization, Doctoral Thesis, University of Illinois at Urbana-Champaign.

5

Common Understanding of HALT Approach is Critical for Success

A new scientific truth does not triumph by convincing its opponents and making them see the light, but rather because its opponents eventually die, and a new generation grows up that is familiar with it.

Max Planck, *Scientific Autobiography*

It must be considered that there is nothing more difficult to carry out nor more doubtful of success nor more dangerous to handle than to initiate a new order of things; for the reformer has enemies in all those who profit by the old order, and only lukewarm defenders in all those who would profit by the new order; this lukewarmness arising partly from the incredulity of mankind who does not truly believe in anything new until they actually have experience of it.

Niccolo Machiavelli, (1469–1527) *The Prince*

Next Generation HALT and HASS: Robust Design of Electronics and Systems, First Edition.
Kirk A. Gray and John J. Paschkewitz.
© 2016 John Wiley & Sons, Ltd. Published 2016 by John Wiley & Sons, Ltd.

5.1 HALT – Now a Very Common Term

HALT has been a common term in reliability for the past decade although it is widely misunderstood by many. Reliability engineers who have not used HALT are not aware of the typically very large range of thermal and mechanical strength of most electronics without moving parts. Lack of knowledge of the empirical strength of electronics then leads to the fear that HALT will result in the wasted effort of chasing failure modes that could not occur in the end-use conditions. It also prevents the discovery that many systems have inherently large stress margins and are already at the limits of the fundamental limits of technology. For those products with large margins, higher levels of stress can be used for HASS processes to find latent defects in manufacturing much quicker.

Many times a weakness found in HALT cannot be easily proven to be a future risk to reliability of the product. Many companies using HALT for the first time have dismissed a weakness found in HALT and only to discover it later as a significant contributor to field failures.

Winning over the hearts and minds of engineers to support the new paradigm of HALT and HASS methods is critical to a company's subsequent success with it. There are several paths to helping the skeptics understand the potential value and major paradigm shift of HALT and HASS methods.

If a company has a good FA (failure analysis) process and has good records of the root causes of field failure mechanisms, it is not difficult to show that the failure mechanism would have been stimulated with a high chance of detection in thermal cycling and vibration. Examples of latent defects found in product field failures that would have a good chance of being found in a HASS process could be loose connector, no solder or a cold solder joint or adjacent components shorting after making contact.

Another strategy for demonstrating the potential value of HALT or HASS is by taking a sample of products that have a known latent defect mechanism or weakness that would be stimulated to failure with HALT or HASS stress and detecting the same latent defect when HALT or HASS is applied. The challenge is that for this to be effective either the latent defect has to be in a large percentage of shipped product or there have to be enough samples of a low rate latent defect to have a reasonable probability of detection in an application of HALT.

5.2 HALT – Change from Failure Prediction to Failure Discovery

The biggest challenge to HALT is a common correct understanding of the shift from probabilistic statistical prediction of an average, MTBF or MTTR, to deterministic discovery of weaknesses that could cause unreliability. Reversing decades of the misdirection of traditional reliability engineering beliefs in the value of reliability predictions for electronic systems, even though there has been no evidence of correlation between reliability predictions and field failures, is a daunting task in most electronics design organizations.

It can take weeks or months to implement and apply the first HALT procedure and even longer to see the benefit in lower rates of field failures. The time between spending the resources and seeing the benefits of HALT is such a precarious period that it helps to have a 'HALT champion' to keep the skeptics at bay. Yet after HALT and HASS have been applied for new product development and the benefits of HALT have been observed first hand, the adoption of HALT methods typically becomes the main focus of a company's reliability development efforts.

HALT is based on a simple concept that 'a chain is only as strong as its weakest link'. An electronic system is a complex chain of interacting components and assemblies. The capability of the system to reliably operate within the variations of the end-use environments is limited by the lowest functional margin or the weakest subassembly or component. A weak link in the design can result in a catastrophic failure, or it could result in a system that has intermittent or marginal performance, making it unreliable in its operation throughout its use period.

HALT is not simply vibration or thermal stressing to catastrophic failure. It is an empirical stress test process in which a single stress or combination of stresses is applied to an electronic or electromechanical system until it fails to operate. The point at which the stress can be said to cause failure may not always be a discontinuous function, such as a shutdown or functional lockup of a digital system, and it may have to be more clearly defined. In the case of a degrading performance or increasing errors, a more specific definition of the point of operational limit or 'failure' should be determined after observing the performance of a system in the first HALT. Thermal HALT applied to digital electronic systems will generally cause operational or 'soft' failure before a destruct limit is reached. In some systems such as an analog audio amplifier, the gain of the system may degrade in a continuous fashion as the thermal stress increases. Because thermal HALT does not typically cause catastrophic failure in digital electronic systems, the operational limits it can be applied many times to isolate the cause of a low thermal margin, or to test alternative suppliers of key components. Vibration HALT and voltage HALT have a much higher risk of causing catastrophic component failures. In planning HALT, the stresses should be applied from the stresses that will have the least risk of a destruct limit, where the test sample would need a component or subsystem to be replaced, to the stresses that have a higher risk of a destruct limit. During product development, samples are typically scarce and in high demand by other departments, and therefore it is important to gather as much stress margin data as possible from each sample used for HALT.

5.2.1 Education on the HALT Paradigm

Implementing a new reliability development paradigm in a company can be a perilous journey. This is especially true with introducing HALT concepts and processes as it represents a significant change in the frame of reference of reliability development. The perspective of HALT and HASS in testing is one of discovering the empirical strength, and not testing based on using models as a basis of quantifying a product's life duration. HALT does not provide a maximum, minimum or mean time to or between failures. HALT is used to find limits and to determine if a product is robust and the design is at or near the FLT. HALT is based on the intrinsic strength and capability already in electronics assemblies, and not simulation of worst case stress conditions, or what the 'average' LCEP may be in the future.

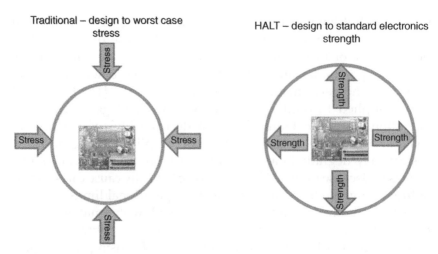

Figure 5.1 The change in orientation for the designed strength of electronics

The drawing in Figure 5.1 illustrate the change from the traditional frame of reference for reliability development of electronics to the frame of reference for HALT.

When the acronym HALT is explained to those unfamiliar with the term as a 'life test' they expect that HALT will provide an annual failure rate or MTBF number, which it does not and cannot, and there is no accelerated testing that can.

The reason that no singular or series of accelerated tests is able to provide a simultaneous field time equivalence of aging is because each stress stimulus applied to electronics assemblies induces fatigue damage to all the many different material bonds and structures at different rates depending on many different factors such as location and orientation on a PWBA, adjacent components, size and weight and material composition among many other factors.

Unreliability in electronic systems can be from complex phenomenological problems that are often accelerated by multiple stresses. The rates of failures for latent defects and normal aging wear-out failures from the cumulative fatigue damage are typically dependent on multiple stresses and their interactions. To know the time to failure for an electronic system it is necessary to know the models and strength distributions for each latent defect or wear-out mechanism, the corresponding internal product response of all of the fatigue producing mechanisms, and the variable field stresses and the distribution of the stress magnitudes and durations to fielded product in different end use environments.

The rates of fatigue damage accumulation become even more difficult to model when considering multiple stresses that are applied simultaneously, as occurs in most end-use conditions, which (especially with mobile electronics) have wide distributions in durations and levels of combinations of environmental stress.

Because of this there has to be a significant change in test goal orientation, that is changing from the goal of projecting life entitlements of electronics to finding weak elements that may impact the life entitlements.

An example of a common electrical failure mechanism is that of an electrical connector developing fretting corrosion. Fretting corrosion is the name for a build-up of insulating, oxidized wear debris that can form when there is small amplitude cyclic motion between electrical contacts. The small cyclic movement comes from both mechanical vibration and the expansion and contraction of thermal cycling. For a tin-plated connector surface, the pressure between the contacts cracks the thin tin oxide surface providing clean tin for the conduction path. Over time the microscopic motion between contact surfaces causes a build-up of tin oxide debris between the two surfaces and results in a high resistance, leading to an high resistance or open conduction path. The rate of fretting corrosion becoming a failure mechanism is dependent on many variables and interacting stress factors such as the number and magnitude of thermal cycles, vibration, contact force, number of connector insertion and removal cycles and humidity. An accelerated test to determine a time to failure for this failure mechanism would not only have to know the field stress conditions and durations, but would also require knowing at what level of series resistance in the connection path, which attenuated the power or signal levels, would cause functional failures. Fretting corrosion failures can be challenging because disconnection and reconnection of the connector mating surfaces removes the insulating debris and the failure mode disappears.

A critical factor for success begins with educating the company's top technical and financial stakeholders on the new paradigm shift of HALT, the reasons for it and how it is performed. A key to having success with HALT, that is the discovery and improvement of stress–strength margins, requires that management, design engineering, test engineering, manufacturing engineering and procurement departments have a common understanding of the big picture of the HALT methods and the goal. Education needs to be done early and with as many key personnel during a common meeting or several meetings in a short time period, before it comes time to actually perform HALT. Otherwise the work of educating each skeptical key player in a serial fashion over time will provide each opposing engineer more opportunity to spread their fear, misinformation and misunderstanding of HALT, which puts the success of HALT at significant risk.

Simultaneous education of key personnel is extremely important to the transition to, the application of and realizing the benefits of HALT methods. Without a common and early education of the many personnel required for successful HALT, implementation can be delayed significantly.

Key personnel that will be required to successfully support a HALT and HASS program are the following:

- Management – The most critical support will be needed at the top levels of management to provide the resources and schedule that will be needed for implementation. The time between allocation of funding for offsite HALT lab use, or the setup of an internal HALT lab, and observing the improved field reliability of products shipped may be a year or more, and management's commitment to the HALT processes will be crucial to reaching the time when field reliability shows the benefits of HALT.
- A cross-functional HALT team – For HALT to be successful, it requires many different engineering functions.
  ○ Team leader – Ideally, the team leader should be a champion of HALT methodology with a good understanding of the root cause of failures in the company's products. The team leader coordinates all the resources necessary for HALT, writes the HALT procedure, directs the HALT and addresses the improvement opportunities discovered in HALT and writes the final HALT report.

- o Software engineers write code or determine what code and functions should be applied during the HALT for the most comprehensive test. They may also help with data collection during HALT. They may also be involved with code changes to improve software timing related thermal operation limits.
- o Mechanical engineers help with subsystem HALT configurations and fixture development when needed. They also help determine improvements to increase mechanical robustness when a weakness is discovered.
- o Electrical design engineers help to determine what electrical test circuits and subsystems should be exercised during HALT, and they help isolate the limiting components in a low margin circuit when found.
- o Test technicians perform the HALT procedure under the direction of the team leader
- o Development engineers provide inputs for special tests that should be applied and data that should be collected and will help with troubleshooting HALT issues for newly designed products.
- o Failure analyst is a critical player in helping isolate the root cause of a low margin or failed component.

- Other departments may also be involved such as
  - o Procurement will be needed to locate alternative component or subsystem suppliers to help improve circuits or systems with low operation or destruct margins.
  - o Logistics will be needed to schedule and coordinate the availability of units for testing and time to complete the HALT.

Imagine that you are your electronic systems company's reliability engineer or you have been involved in reliability qualification or validation testing of its products for several years. You have experienced field failures that resulted from design margin issues that were overlooked during the development process, as well as some from mistakes in manufacturing. Reliability development in your company consists of running tests that simulate the estimated LCEP, or design engineers applying limited stress to their own predefined 'that's good enough' level, or estimates on what may be the worst case stress environmental conditions for the product.

You have just learned the generic methodology and some of the benefits of HALT from reading a book on the subject, attending a class or listening to a webinar. Now that you have heard of the process and benefits you may want to try HALT for your company's new product. You locate a nearby test lab that has a HALT chamber so that you can go and do a HALT. You just got funding and new product prototype samples to use for HALT, so the next step is to take them to the lab and starting finding its stress limits, or is it?

5.3 Serial Education of HALT May Increase Fear, Uncertainty and Doubt

Let's continue with this story of typical introduction of HALT methods scenario to illustrate the challenges.

You find an outside test lab that can perform HALT a few miles away from your company. You have five samples of the new product, support equipment to operate and monitor the UUTs, and possibly a technician or even better you have a design engineer for the product you are going to test to go to the lab with you. The environmental design specifications for the product are 0°C to 35°C.

The unit to be tested has its over temperature protection circuits defeated so that the raw thermal performance of the product can be discovered. The lab personnel help you to set up the first sample of the product in the HALT chamber. You begin the thermal portion of HALT to find the lower temperature operational limit and the upper temperature operational limit. Since the product in this case is a microprocessor based digital system the operational limit is found, but no destruct level is found. In the five samples used for HALT, you find upper temperature empirical operational limits at 70, 72, 90, 117 and 110°C. The lower temperature operational limits for the five samples are found to be −55, −45, −50, −58 and −47°C.

Since all samples are operational after the thermal HALT and when inspected in detail have no obvious thermal damage they can all be used for the vibration portion of HALT. The final stress used in HALT is vibration and two of the samples fail when the vibration level reaches the maximum vibration level of the HALT chamber. The failure mechanism on both is a broken lead of a capacitor mounted high off the PWBA. You and the design engineer repair the capacitor leg and use an adhesive to attach it to the PWBA. To verify the HALT improvement, you apply HALT vibration to the same maximum level and the glued capacitors do not fail.

After you complete HALT at the outside lab and come back to your company, you wonder why there is approximately a 40°C difference in upper temperature operational limits between the five samples. You realize that wide variation in limits may be an indicator of some components' inconsistent manufacturing processes, or significant sensitivity to inherent parametric variations of a component, which if it increases its variation could significantly impact field reliability. You hope that you can have the manager of design engineering support an investigation into the cause of the wide upper thermal limit variations between the samples. When you meet with him he tells you that his department is very busy with the next design and his limited resources will not be available for the following reasons:

- The product meets the design specifications, and even the worst sample has 35°C margin above design specifications.
- The product will never see 70°C in its worst case use; therefore if it does fail it's the customer's fault.
- We do not have time to redesign the product to meet your HALT stress requirements.

How do you address these obstacles from an engineering manager for resources needed to identify the weaknesses and potentially improve the product robustness and reliability?

Let's say you spend an hour with the design manager, overcome his objections and get help and support to continue from the design engineers. With the help of two design engineers you isolate and determine that a 10 W IGBT (insulated-gate bipolar transistor) is the most likely cause of the upper temperature operational limit. Fortunately you find a 20 W IGBT in the same size package and voltage and use it to replace the 10 W IGBT. You go the HALT lab two weeks after your first HALT and find that all three new samples have an upper operational limit above 115°C and again no thermal destruct limit is found.

5.3.1 While You Were Busy in the Lab

Later in the week you find that during the time you have been doing the HALT at the local environment test lab, a skeptical design engineer (you have yet to speak with) has heard that you want to 'over-design the product for stresses it will never see' and has spoken to others in the design and procurement departments. Now you start hearing that engineers you have not yet spoken with comment on your "desire to add unnecessary product cost and to overdesign" a product for an environmental stress well beyond end-use conditions. You find that you are teaching the HALT paradigm shift to each skeptic as you find them, convincing some, but the ones you have not spoken with are also spreading the fear of costly over-design because of their misunderstanding of the HALT methodology.

When you go to the purchasing department you find out that the higher wattage IGBT will add 50 cents of additional cost to a product that retails at $700.00. Now the vice-president of engineering hears about the additional product costs if the IGBTs are changed. Since it will reduce the profit margin on an already competitively priced product, the VP then asks very similar questions to those that the design engineering manager asked previously. The design manager attempts to explain the reasoning in their half-hour meeting with the VP, but lacking the experience of the new paradigm of HALT and supporting data they don't succeed in convincing the skeptical VP to accept the change to a higher power component.

5.3.2 Product Launch Time – Too Late, But Now You May Get the Field Failure Data

Ultimately, increasing the wattage of the thermal operation limiting IGBT is not implemented and the product is released to market. In a year or two years, you may be able accumulate the warranty return data and find that IGBT failures have been a significant field failure contributor. You may be fortunate enough then to have an engineering change order to replace the original IGBT to the higher powered IGBT and observe the change in rates of field failure. Unfortunately, a more likely scenario would be that after the product begins, a new design engineering manager has just joined the company, or has moved from another division during reorganization only a few months ago.

Unfortunately they are not familiar with the HALT methods. So, back to square one in teaching HALT to the new design engineering manager, and then possibly many new design engineers, reliability engineers and department managers since you first began your path to introducing HALT at your company months or years ago.

Unfortunately the personnel and management can change rapidly in today's technology companies with reorganizations, mergers and consolidations. During the many months after HALT there is a good chance that most skeptics that you tried to convince about the benefits of the HALT paradigm will just remember the efforts and costs to set up and apply the first half of the HALT process. They may not recall that the weakness was not mitigated or may not be aware that a weakness identified was later found to be the cause of a percentage of warranty returns months later. In the future, the skeptics who have watched the application of the HALT weakness discovery process, but not the corrective action, are likely to state that 'We did HALT but found nothing' or 'HALT was a waste of time: the only issues found were in conditions that would not exist in its end-use.' You now have to address those who think that HALT was used despite the fact that it was only started, and not finished. Or are you still a reliability engineer at the same company?

6

The Fundamentals of HALT

6.1 Discovering System Stress Limits

Unlike most testing in product development, HALT has no predetermined pass/fail criterion or temperature requirement to achieve. It is an open ended weakness discovery and strength improvement opportunity process. The HALT will be completed when the failure and limit analysis determines that the product's margins cannot be improved with the current technology or within the costs constraints.

The purpose of HALT is to find weaknesses and improve stress margins to the fundamental limit of the technologies so that the product is as robust as current standard materials and processes allow. The goal is to increase the limit of stress margin to the FLT. The FLT is the point at which the margin cannot be increased without resorting to non-standard materials and methods. Performing HALT on similar products developed over time will result in a general understanding and benchmarks of what most similar assemblies, and systems temperature and vibration limits are with current technology and what stress limits should be expected.

There are times that the hardware being tested in HALT is already very robust and is at the FLT. If the product is at the FLT, no changes

Next Generation HALT and HASS: Robust Design of Electronics and Systems, First Edition.
Kirk A. Gray and John J. Paschkewitz.
© 2016 John Wiley & Sons, Ltd. Published 2016 by John Wiley & Sons, Ltd.

are required to improve the design margins, and a HASS process, if used, can be developed quickly.

6.2 HALT is a Simple Concept – Adaptation is the Challenge

HALT is an inherently simple concept in theory but requires careful consideration and adaptations to the product during application and failure analysis. Almost all environmental test processes used for design validation in conventional reliability development can also be used for HALT applications. HALT is fundamentally the application of any stress to empirical operational limit, destruct limit or test chamber/equipment capability limits for the discovery of relevant weak links in a design.

The terminology of HALT is sometimes misunderstood. HALT is not a 'life' test at all, as there are no appropriate acceleration algorithms and no acceleration factor. HALT cannot provide information regarding the expected life of the product in the field. There has been no evidence presented that any accelerated life tests can quantify the life entitlement of a complex electronic system, despite the belief by many engineers that there are. This is because there are so many potential physical failure mechanisms in an electronic system, each with unique rates of fatigue damage or chemical reactions from the same stress condition depending on many complex interacting factors.

For many electronic systems there is a wide distribution of life cycle stress and end use operating conditions for which only an average value is assumed. The life cycle stresses for cell phones is a good example of wide distributions of life cycle environmental conditions. The distribution of stresses that a cell phone is subjected to during its end-use extends from low stress to high stress abuse, but for the end user the difference between normal use and abuse is not easily defined. Many cell phone and other portable consumer electronics producers specify the maximum operating temperature as 35°C, so if the user operates it at human body temperature or outdoors on a hot summer day of 37°C, they are exceeding the manufacturer's use specifications. Of course, most users are probably unaware that they regularly exceed cell phone use temperature specifications if the cell phone is at human body temperature, and the users and market would not accept 35°C to be the actual maximum environmental operating temperature.

The weakness found with one stress in HALT may not be the same stress that causes the failure mechanism to progress to failure, as many failure mechanisms can be driven to failure by different stresses. An example of this would be a crack in a conductive trace, or plated through hole or via between board layers. Both thermal cycling and mechanical vibration will accelerate a crack in a conductor but at different rates. Which of the field stresses or combination of stresses ultimately results in an open conduction path is dependent on cycles of end use environmental stresses. If the conductor crack is precipitated to an observable failure in HALT during vibration it is a valid stress for finding a potential reliability weakness despite the fact that there is little or no vibration in its end use.

6.3 Cost of Reliable vs Unreliable Design

HALT helps discover weaknesses in a design in the first stages of hardware development. In most cases only one or two changes are needed to a hardware design to significantly increase stress capability.

Some changes to a design require no change in hardware at all. In Allied Telesis 'Software Fault Isolation Using HALT and HASS' that is reprinted in Appendix A.2, Allied Telesis found a software fault during cold thermal HALT that was caused by a reset pulse to a programmable logic device (PLD) being too short in duration. The fault was observed at −10°C. After the PLD code was updated, the lower temperature operational margin was increased to −50°C. This increase in low temperature margin, a change of 40°C, only a change in PLD code.

In the same paper, they were able to adjust the tuning of a switch to improve the thermal operational limits from −20°C to 70°C up to −60°C to 100°C. Other changes that the author has observed that can lead to large gains in thermal stress operational limits are changing a resistor value, and changing a ⅛ watt diode to a ¼ watt diode that increased the upper operational temperature limit by 30°C.

Figure 2.3 (a) and (b) of the paper 'Reliability Prediction – Continued Reliance on a Misleading Approach' shows two examples of how improving vibration capability and robustness can be achieved by adding additional standoff supports to a circuit board to reduce the flexing of a board, or changes in the location of a component. All changes to an initial design to improve its robustness have some costs. The sooner the change is made in the development, the lower will be the total costs for the change. Some margins may have more costs to improve, but without finding the margins by stressing to empirical limits in HALT there is no discovery what they are and therefore no opportunity to improve them.

6.4 HALT Stress Limits and Estimates of Failure Rates

Figure 2.18, the chart of the Cisco Router line circuit warranty returns versus thermal margins, shows a probabilistic relationship between thermal margin and percentage of returns. If the design, materials,and manufacturing processes as well as end-use LCEP are similar, and field failure rates and HALT limits of predecessor systems are known it may be possible to show a comparative probability of failure rates based on thermal operational margins. If all other critical factors contributing to the system's reliability are the same, there will likely be a correlation between higher operational margins and reduced warranty return rates. For successor systems that have similar genesis and use conditions, the previous comparisons between stress margins and historic return rates may provide a business case for making a change to increase a design's stress margins against the probabilistic savings in warranty returns.

The relationship of increased thermal margins and the reduction of warranty returns should be assessed with each product to help establish a benchmark of fundamental limits for the technologies being used, and to determine the point of diminishing returns on increasing thermal margins and reducing warranty returns.

6.4.1 What Level of Assembly Should HALT be Applied?

HALT is a process to discover the weaknesses in the function of a combination of critical elements, such as those found on a circuit card assembly. HALT was not intended to be applied to individual components or discrete devices operating alone. Reliability testing of individual components can be more focused on the specific device parametrics and limited device specific mechanisms based on the physics of failure. The main goal of HALT is to find the weak link in a operational assembly of components, materials, electrical parameters. It is a comprehensive test of system interactions and operational strength of a system.

In most cases circuits or subsystems with only a few active components should included in a HALT with the larger system. If a circuit card with a small number of components is found to be a cause of low margin, HALT methods can be used on it to isolate the limiting component in that circuit assembly.

6.4.2 HALT of Supplier Subsystems

Many electronic systems are developed with subsystems from many suppliers. Power supplies, disk drives and other peripheral systems that are available before the prototypes of the whole system are available. Power conversion subsystems (i.e., power supplies) are typically available early in the hardware build phase. Power supply operation is critical to system reliability. HALT is very useful in evaluating power supplies because they can generally be operated independent of a system using passive or active loading automated test equipment (ATE) systems. Because power supplies generally have higher mass components such as transformers and large capacitors, there is a higher probability of finding mechanical weakness such as components spacing being insufficient resulting in electrical shorts, or weaknesses in component mounting being found in vibration HALT.

If these same systems have been used in predecessor products, have a field history of reliability and have no changes to their designs, there may not be much benefit from separately applying HALT or HASS to the subsystem. The subsystems should be included in HALT of the higher level system, if feasible, because operational weaknesses may be discovered in the interface between the known reliable subsystem and a new system design.

6.5 Defining Operational Limit and Destruct Limits

Many times the functional limits discovered in thermal HALT are easily observed by a sudden system lockup, which is typical of digital systems during thermal HALT, or a destruct limit or 'hard' failure is discovered in which there is a discontinuity in operation. Thermal HALT in analog systems such as an audio amplifier may lead to having a continuous performance degradation of the output such as distortion or gain.

In cases where the performance of an electronics system degrades, the stress operating limit may need to be defined based on the level of its operation. An example would be an audio amplifier in which the gain decreases to near zero gain at some high temperature. In this case the upper operational temperature limit may be defined in a HALT evaluation as the point at which the amplifier gain decreases to 10% of the gain at room temperature. This definition of the stress operating limit should be used for all samples subjected to HALT under the same inputs, in order to compare and look for wide deviations in performance and function of the product being subjected to HALT. Wide performance deviations can be an indicator of poor process control of a supply chain manufacturing process.

6.6 Efficient Cooling and Heating in HALT

The largest thermal mass in a HALT chamber during a HALT is usually the chamber walls and vibration table. HALT chambers typically have internal air ducts that can direct the airflow to specific areas of the UUT using flexible aluminum ducting. Directing the high rate airflow to the UUT provides the highest rate of UUT thermal change and reduces the amount of thermal energy used for HALT and HASS. Directing the airflow to the UUT also reduces the heating and cooling of the chamber walls and vibration table. A typical HALT configuration of a circuit board is shown in Figure 6.1.

6.6.1 Stress Monitoring Instrumentation

There are several other instruments that should be available for performing a HALT and the setup of a HASS process. Some of the instruments for monitoring the stress response of the product, such as thermocouples and data logging, or vibration spectral density analyzers are optional systems that can be included with a HALT chamber system or acquired independently. For thermal HALT, thermocouples and a data logger should be used to record temperatures of UUT operational limits during HALT and to ensure that the UUT has reached the thermal set points.

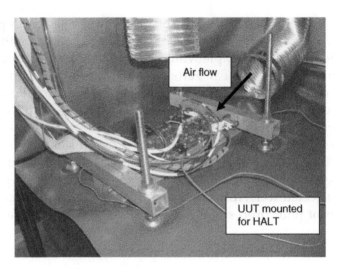

Figure 6.1 Typical mounting of a circuit board with aluminum ducts directing air flow across the UUT. Source: Felkins, 2013. Reproduced with permission of Charles Felkins

A spectrum analyzer is used with accelerometers for observing and recording the vibration power spectral density (PSD) and gravity root mean squared (Grms) response of the product for the measurement of product response from vibration. The accelerometers should be low mass type (e.g. 4g), with frequency of $\pm500g$ (standard acceleration of free fall: $g = 9.8\,\mathrm{m/s^2}$). The accelerometers should be small enough to be mounted in a central location in the UUT during HALT, and light enough that their mass does not significantly impact or alter the vibration dynamic characteristics of the UUT. Examples of both single axis and triaxial low mass accelerometers are shown in Figure 6.2.

6.6.2 Single and Combined Stresses

In introducing HALT processes to a company it is best to start with the traditional HALT process of applying one single stress at a time, and finding each limit of that particular stress for each sample used for HALT. Using a single stress in HALT can make it easier to determine the root cause of the operation and destruct limits.

Figure 6.2 Low mass triaxial and single axis accelerometers. Source: Dytran Instruments, Inc., 2013. Reproduced with permission of Dytran Instruments, Inc

Stresses of temperature, vibration, voltage and others can be and should be combined for additional HALT limit data after the operational limits of a single stress are discovered. New units, if available, should be used for combinations of thermal and vibration HALT. Simultaneously applying thermal and vibration stress has a synergistic effect for creating fatigue damage. Changing the temperature of an PWBA shifts the natural vibration resonance frequencies and therefore the mechanical fatigue damage distribution during HALT.

An illustration of the effects of temperature on the natural resonance frequency of a FR4 circuit board substrate is shown in Figure 6.3. Two of the PSD curves shown are from measuring the z-axis vibration level at the center of the circuit board at two temperatures. It can be observed that the vibration resonant peak frequency shift of a PCB shifts from 35 Hz at 70°C to 48 Hz at at −35°C . Variations in the end use conditions and the variations in manufacturing strength will produce a distribution of end-use shock and vibration responses, and combined stresses increase the probability of discovering weaknesses that lead to field failures.

Experienced HALT users may run combinations of stresses to better compress the HALT evaluation time and then use individual stresses to isolate the cause of a low stress limit. Some stress combinations that

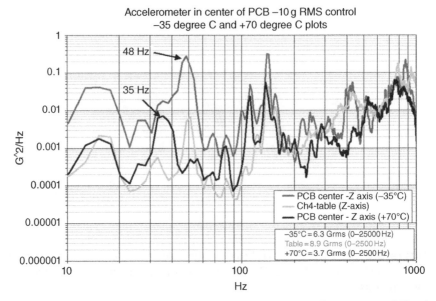

Figure 6.3 PSD showing peak vibration resonant frequency shift of PCB at −35°C and 70°C. Courtesy of CALCE – University of Maryland.

are very interdependent, such as thermal stress combined with voltage and clock frequency stress when applied to digital electronics systems, can result in more valuable variable performance data compared to attribute data (operational, not operational). Variable data from combined stress HALT can provide higher resolution of margin comparisons between samples for discrimination of timing and signal integrity weaknesses.

6.7 Applying HALT

Before closing the environmental stress chamber to begin HALT all units should be verified to be operational. Each unit should have the functional test routines that will be applied during HALT before beginning the stress steps. This is to save time in HALT by removing any samples that are DOA (dead on arrival) before closing the chamber and applying stress.

6.7.1 Order of HALT Stress Application

HALT is performed using the order of stresses starting from the least likely to highest probability of reaching a destruct level. This order is used to gain the maximum limit information from each sample before it becomes non-functioning. The sequence of stresses in HALT is typically low temperature, then high temperature and then vibration. HALT is not limited to these basic stresses only. Depending on the system and its functions, advanced HALT users will apply many more stresses, such as voltage and clock frequency to determine empirical limits. Once limits of individual stresses are found, combinations of stress stimuli, such as temperature and voltage, can be used to create multidimensional empirical boundary maps as shown in Figure 4.20.

After all of the singular stress operational or destruct limits are found, combinations of stresses should then be applied to observe if interaction between stresses changes the limits found during singular stresses. Applying combinations of stresses, such as temperature and voltage, may help with isolating the cause of a discovered design weakness.

Starting a HALT with combined stresses should only be done after having a lot of experience with HALT and on a similar type product.

Typically new samples are used for each step, but many of the units used for thermal HALT can be used for vibration HALT, as most thermal HALT evaluations find the thermal operational limits only and do not result in significant damage or destruction.

Random vibration from repetitive pneumatic shock produces rapid cumulative mechanical fatigue damage in electronics and electromechanical systems. Because of the more rapid accumulation of fatigue damage from vibration, the vibration HALT is much more likely to cause a destruct (catastrophic) failure. Frequently in vibration HALT an operational limit is also simultaneously a destruct limit. A vibration HALT can stimulate a solder crack to a 100% fracture of a solder joint, and the mechanical displacement while applying vibration to the UUT makes the fracture an open circuit, and the fault detectable when monitored during HALT stress. When the vibration stress is removed the fracture solder joint makes contact and closes, so the fault disappears. Although this appears to be an 'operational limit' it is both the operational limit and the destruct limit since the solder joint ultimately needs repair, not just a power cycle or reset of the system to remove the fault. An example of a simultaneous

Figure 6.4 SEM picture of the BGA solder joint shows cracks on the top with some connection. Source: CALCE – University of Maryland. Reproduced with permission of CALCE – University of Maryland

operation and destruct limit would be a crack in a trace or solder ball on a BGA, where vibration provides mechanical displacement for complete separation or open circuit that can be detected during circuit operation. An example of such a fracture across a BGA is shown in Figure 6.4.

The vibration induces mechanical displacement to cause the circuit to become open, and when the vibration stress is removed the crack closes along with the circuit to make the system operational again. This also illustrates the importance of functional monitoring throughout the HALT process. So when the stress is removed the location of failure may be hidden and difficult to find and, without root cause failure analysis, may be misinterpreted as an operational limit only and not the destruct limit that it is.

6.8 Thermal HALT Process

The traditional thermal HALT protocol is to first find the thermal LOL (lower [temperature] operating limit) and then the thermal UOL (upper [temperature] operating limit) and then the LDL (lower destruct limit) and then the UDL (upper destruct limit). Thermal LOL and UOL on digital systems are defined by a 'soft' failure. Soft failures in a digital system are failure modes such that when the stress is removed and the system is reset it will operate normally. It has been the author's experience with many HALTs applied to digital IT hardware systems that it is very rare that thermal HALT will result in a catastrophic or 'hard' failure.

Since thermal HALT on digital systems results in ceasing the operation of the system, any higher or lower temperatures will be applied to non-operational systems. There may be in some cases good reasons to apply stress in HALT on digital or other systems beyond operational limits, but in most cases destruction limits of the system will occur most likely when the components or materials reach a change in state as in plastics melting or solder reflowing. In most cases, a material change of state is a failure mode that is not a potential cause of field failures and therefore an irrelevant failure mode.

6.8.1 Disabling Thermal Overstress Protection Circuits

Many electronic circuits have over temperature protection (OTP) included in a design to protect the system from damage or fire due to overheating of a circuit or component. Sometimes an OTP is designed into active components to turn them off before they become operationally unstable. Most CPUs used in personal computers and other hardware have the ability to decrease their processing function or to shut down based on their internal temperature monitoring circuit.

Since the goal of HALT is to discover the raw empirical operational margins and not the designed-in limits, devices or circuits that are designed to shut down at high temperature to prevent fire or damage during normal use need to be defeated. HALT may be used to validate the operation of an OTP but after confirmation or observation of the deviations of the temperature of activation of the OTP across several samples, the OTP should be disabled to discover the true high temperature HALT operational limit. If the HALT UOL also turns out to be a UDL and those limits are close to the OTP limit, there is a risk of the system being catastrophically damaged and therefore negating the intended benefit of the OTP.

Where the OTP is internal to an active device, as it is with a CPU, disabling it is not generally possible. If a component with an internal OTP is in a system, then using external cooling focused on the component is the next best option. A flow of air across the device from a hose, or attaching an externally liquid cooled heat sink to the component can prevent it from reaching the OTP activation temperature, and high temperature HALT can be continued on the system.

6.8.2 HALT Limit Comparisons

Airflow in a HALT chamber is at significantly higher velocities than in typical thermal environmental chambers. The high airflow rates in HALT chambers may result in a difference in the thermal operational limits of the system compared with standard thermal chambers. The air velocity in a HALT chamber is very high to provide rapid thermal transitions and will in most cases result in smaller thermal operational differences between components compared to the thermal conditions of the final system configurations. The components may actually operate at a cooler temperature during HALT than in its final assembly configuration in end use at the same air temperature. It is important to always remember that HALT is not a **simulation** of end-use conditions, but instead a **stimulation** of the product to find weaknesses.

How relevant and significant the weakness is to field reliability can only be determined after a weakness has been discovered and analyzed. In some cases, the maximum stress levels produced by a HALT chamber, especially with vibration stresses, do not produce a failure during a short application of each HALT step. If this occurs with all samples of the product in vibration HALT, then the duration of operation at the highest level of stress the chamber is able to produce can provide the variable data that should be compared for determination of strength limits.

Reaching the stress limits of the chamber is a good reason to end a HALT as it would be a good indication that the product is near or at the FLT. Otherwise to arbitrarily stop HALT at a stress level that someone has determined as being 'good enough', can prevent finding a very relevant weakness that might be very close to that arbitrary limit, and that would only need a small change to improve it.

Some HALT chamber manufacturers control HALT and HASS temperature with a two-channel input control system. Rapid thermal transitions of the UUT are produced by 'overdriving' the air temperature above the set desired product temperature. One channel monitors the product temperature and another channel monitors the air temperature so that high thermal transition rates can be forced on the UUT. As the product reaches the desired set point, the difference between air temperature and product temperature declines until air and product temperatures are the same.

For HALT chambers with two channel control the location of the product temperature is important for proper control. The product temperature monitoring thermocouple should not be located deep in the UUT, or on a large thermal mass, such as a transformer, or on or very near a heat generating component. If the thermal lag on the product reference measurement point is far behind the air temperature, the air temperature forcing function could exceed material properties on another lower mass portion of the UUT and cause damage such as solder melting, which is an irrelevant failure mode in HALT.

How much the difference between product and air temperatures is allowed can be set in the control settings. The actual temperature thermal transition rate is dependent on many factors of the UUT. Factors are the thermal mass of the UUT, the placement and direction of airflow on the UUT, the location of the UUT temperature monitoring thermocouple and the heat loads generated by power dissipation during the operation of the UUT being tested.

There are many benefits in applying higher thermal rates of change besides reducing the time it takes to perform thermal HALT. Higher airflow in HALT chambers provides rapid thermal transitions and across the test samples creates both spatial and temporal thermal gradients during thermal cycling. Larger gradients produce higher thermomechanical shear and strain stresses between material bonds and interfaces due to differential thermal coefficients of expansion between materials. Thermal gradients also differentially skew the component and material parametric properties.

The higher thermal rates of change produced by a typical HALT chamber relative to a traditional mechanically cooled chamber are the reasons that HALT/HASS chambers are capable of finding weak bonds and material interfaces in a few short thermal cycles, along

Figure 6.5 Typical HALT chamber. Source: QualMark 2015. Reproduced with permission of QualMark Corporation

with the synergistic benefit of random multi-axis vibration when combined stresses are applied. A typical HALT chamber is shown in Figure 6.5.

6.8.3 Cold Thermal HALT

For each HALT a minimum of three units should be used to start. The more samples used for HALT the better. More HALT samples gives more confidence in the confirmation of limits and higher resolution of comparisons of sample-to-sample variation of each HALT limit. More samples are better and the number of units needed for any limit (failure) evaluation may change in order to have more specific investigations and isolation of a limit's root cause. If the samples are completely operational after each HALT stress application they can be and should be used for all stresses in the same sequence of application.

Cold thermal HALT will begin at ambient temperature, usually around 23°C. The temperature is lowered to the first cold stress level, which may be within the specifications. The product samples should have thermocouples to monitor the actual temperature of the UUT at each temperature step. There is no hard and fast rule regarding the thermal dwell times at each step, but usually a 10 minute dwell time after the product has reached or is close to thermal equilibrium. A dwell time should be held long enough to apply all functional tests, including power cycling if possible. After reaching each temperature step and during the dwell times, run all diagnostic or system operational stresses to confirm that the system operates with no functional problems. Record all operational parameters and thermocouple readings from the UUT at each step then decrease the temperature by 10°C and repeat the verification of operation.

Power cycling is very beneficial for weakness discoveries, as low operation margins may be exposed by the stress of voltage transients and current surges during temperature stress. Powering the UUT during the transitions down in temperature reduces the heat load during the steps down in temperature and allows some internal cooling of components before rapidly heating when the power is switched on.

Perform all functional tests, power cycling and other test routines that ensure the operation of as much of the system or subsystem as possible. Continue thermal steps down in temperature until the UUT fails to operate or to the thermal limit of the HALT chamber is reached.

A graph of the decreasing temperature HALT profile and the thermal lag of the product response is shown in Figure 6.6. The thermal response of the UUT will be dependent on its thermal mass and its self heating from power dissipation.

The use of liquid nitrogen as a cooling medium results in very dry air in a HALT chamber during cold HALT so moisture condensation rarely occurs. Always avoid opening the chamber at cold temperatures to prevent rapid moisture freezing or condensing on the chamber walls and table.

After finding the lowest temperature functional limit return the UUT to ambient room temperature (23°C). If the UUT is fully functional then record the low temperature limit as the LOL (lower operational limit). If the UUT is not fully functional at ambient conditions then the temperature limit is recorded as the LDL (lower destruct limit) for the

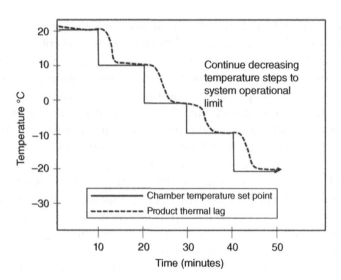

Figure 6.6 Decreasing temperature HALT profile and thermal lag of UUT

sample. If the LOL has been found, the sample should be used for the hot thermal HALT. It is good practice in HALT to perform the same sequence of HALT stresses for each sample, creating the same sequence of cumulative fatigue damage for each sample.

If the thermal LDL has been reached remove the sample from the HALT chamber and replace it with the next sample. Use the same locations for attachment of thermocouples to record each and all samples' product temperature responses in HALT. Use the same product configuration, with the same orientation of air flow in the chamber for each sample.

Make a detailed inspection of the unit to find any obvious component damage or discoloration. After all the samples have been through the first iteration of the thermal HALT procedure, the units that have hard failures because they reached the LDL should have detailed failure analysis to uncover the root cause of the LDL.

6.8.4 Hot Thermal HALT

The same test and measurement configuration for cold thermal HALT should be used for hot thermal HALT. Beginning at 30°C, increase the temperature in 10°C steps, holding each step for 10 minutes or until the thermocouples indicate that the temperature has stabilized. When the temperature is stable, record all thermocouple temperatures to compare between multiple samples subjected to HALT. Perform all functional tests, power cycling and other test routines that ensure the operation of as much of the system or subsystem as possible.

Continue 10°C steps until the UUT fails to operate. Record the temperatures of the UUT at the point of failure. Reduce the chamber temperature to room temperature conditions (23°C) and verify whether the failure was the operational limit where the UUT operates with the stress removed, or a destruct limit or 'hard' failure requiring repair of the hardware to make it operational.

6.8.5 Post Thermal HALT

Samples that have a UOL that is recoverable and is operational after the thermal HALT is complete should be carefully inspected to see if there is any change in components' appearance such as bulging of aluminum electrolytic capacitors or discoloration of the circuit board from a component's high temperature.

In some cases it may be that the product's design is near or at the FLT and that the operational stress margins cannot be improved with any changes using standard materials and processes. If there is a significant deviation of limits between samples of the same product, the root cause of the wide limit deviations should be investigated and isolated. Determining the FLT for stress margins can be derived from knowing the material specifications and from benchmarks of operational limits from HALT of predecessor products plus good engineering knowledge. After performing many HALT evaluations, a HALT practitioner will acquire benchmarks of what level of stresses different types of assemblies should be capable of.

A graph of the increasing temperature HALT profile and the thermal lag of the product response is shown in Figure 6.7.

Decisions on whether to make any changes to the product to improve operation and destruct margins will be made through team evaluation and consensus. It will be based on what cost and effort is required to increase the stress margins. Changing the wattage of a limiting component or a software code change can add significant thermal margins at very little costs if caught early in the design.

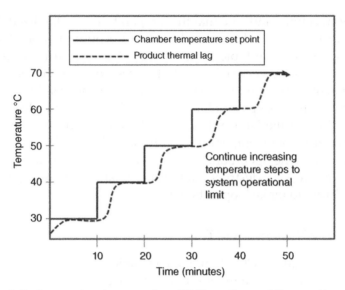

Figure 6.7   Increasing temperature HALT profile and thermal lag of UUT

6.9 Random Vibration HALT

Random vibration is measured in gravity root mean squared (Grms), the square root of the area under the acceleration spectral density (ASD) curve in the frequency domain. The level of repetitive shock (RS) vibration on a HALT vibration table is controlled and measured in Grms. The ASD or power spectral density (PSD) bandwidth in an RS HALT vibration system can extend from 20 Hz to 10 kHz. Unlike an electrodynamic shaker, the shape or frequency spectrum of the ASD input produced cannot be directly controlled in a typical multi-axis RS HALT vibration chamber, only its level of intensity expressed in Grms.

Lack of frequency spectrum control is not necessarily a problem in the application of vibration HALT. According to what has become known as 'Papoulis' Rule', it can be stated that *when a broadband random signal of almost any probability distribution is the input to a narrow band pass filter, the probability distribution of the filter's output approaches Gaussian.* Therefore mechanical fatigue damage is only created in a test object when it is stimulated into a state of self resonance. The UUT acts as its own tuning fork or narrowband filter under vibration. Those frequencies that do not contribute to inducing self resonance do not induce fatigue damage. It is possible to change the attenuation of the vibration frequencies through the use of damping materials such as rubber sheets between the UUT and vibration table, but it is difficult to get a specific vibration spectrum as can be done on a classical electrodynamic (ED) shaker.

High shock pulse peaks produce the highest rates of fatigue damage. Classical ED shaker systems are generally limited to ±3 sigma peak accelerations, to prevent damage to the shaker. HALT vibration from pneumatic hammers is sometimes referred to as repetitive shock (RS) systems. RS systems provide multi-directional high pulse shock peaks that can be as high as ±10 sigma peak accelerations resulting in a higher rate of fatigue damage than an electrodynamic shaker system. Another benefit of RS over ED is that most EDs produce vibration in one axis, as ED shakers with two or three axis rapidly drives up the costs of equipment. RS systems are generally multiple axis which can simultaneously stimulate six degrees of freedom, with three along the x, y and z axes, and three rotational about x, y and z axes. Because of this, HALT vibration is sometimes referred to as 6 DOF (six degrees of freedom)

Generally RS vibration in most HALT systems is highest in the z axis, which is the axis normal to the plane of the vibration table. The vibration level in RS systems is typically controlled by feedback from an accelerometer under the center of the table. Vibration is not uniform across an RS system, and the variations in vibration intensity across the table should be measured before HALT to be aware of where the highs and lows of the variations occur. RS hammer positions may also be changed, which will help balance the intensity across the table, but any changes to hammer positions should be with the guidance of the chamber manufacturer.

The weight of the UUT and the vibration fixtures will affect the levels and frequency spectrum. Because of the vibration fixtures, the weight and locations of the UUT on the table can all change the input vibration levels and spectrum. Therefore the reference level for each sample's stress limits in HALT should be referenced from vibration response measurement of the UUT using the same accelerometer location on each product sample.

All vibration fixtures should be kept as light as possible, while providing sufficient rigidity to support and provide essentially equal vibration energy levels throughout the product. The fixtures should not have any resonances within the desired vibration frequency spectrum (approximately 30 Hz to 3 kHz).

The least fixture is the best fixture. Most HALT vibration tables will have threaded 3/8 inch (10 mm) holes evenly distributed across the table. The simplest fixtures for HALT vibration can be made with 3/8 inch 'all thread' metal rods and a cross bar to hold the UUT down on the table. The vibration table can be a significant heat sink for the UUT. Chamber manufacturers typically cover the table with an insulating material to reduce the thermal coupling of the UUT and table. Rigid blocks, such as solid aluminum, can be placed between the UUT and the table to provide space for air flow under the UUT.

Since the rates of cumulative fatigue damage from thermal step stress HALT are very low, the size of each step and dwell time for thermal HALT is not too critical. When using thermal cycling as a HALT stress, the thermal transitions and dwell time will affect the rate of fatigue damage during mechanical expansion and contraction of materials. Because of the high rate of cumulative fatigue damage during vibration HALT, each sample in should have the same Grms step levels and duration. Vibration has a very short delay between the chamber set point and the UUT, and the vibration dwell time is typically 10 minutes for each step, but that may be extended to complete operational verification.

A vibration HALT profile is shown in Figure 6.8.

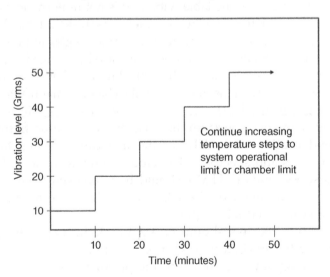

Figure 6.8 Vibration HALT profile

6.10 Product Configurations for HALT

The process of applying HALT starting at the lowest functional assemblies that can be operated and then building up to applying HALT to the full system is usually the best approach. Understanding each subsystem's HALT stimulus limits and knowing the strength of each 'link in the chain' helps isolate the subsystem with the lowest limits or the 'weakest link'. Consider the scenario of testing multiple subsystems and finding significant margins when operated separately but the margins drop significantly when connected to the larger system. It could be a timing issue, voltage drop or signal integrity issue which may be more elusive to isolate, but it points to the interface between subsystems for more investigation into the cause.

HALT on subsystems is best done when connected to the higher level systems located outside the HALT chamber conditions. It may be necessary to fabricate extended cables. In high speed digital systems extensions of signal cables may not be possible because of signal timing issues. In the case of not being able to extend a subsystem or power supply outside of the HALT chamber a small isolation box, or chamber inside a chamber, can be constructed to keep a subsystem at operational temperature while thermal stress is being applied to the UUT. Figure 6.9 illustrates this concept of thermally isolating a UUT inside a HALT chamber while exposing the circuit boards to thermal stress.

HALT may be performed on a subsystem while connected to an ATE functional tester outside the chamber to operate the UUT. The ATE system should be able to activate and exercise the circuit and indicate fault or failure conditions. The advantage of using an ATE system may come from diagnostics and monitoring for better isolation of the cause of a low margin condition and measurements of parametrics before an operational limit occurs. Dedicated functional test systems may also provide better diagnostics on isolating the location of the fault or a component failure.

Many times with high speed digital hardware the bus speeds and signal timing requirements make it difficult or not possible to extend cables from a subsystem to a system outside the HALT chamber. After discovering a thermal HALT limit in a subsystem that is integral to the operation of the larger system and where the subsystem cannot be

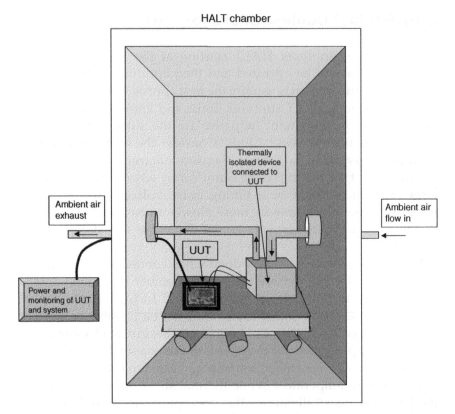

Figure 6.9 A thermal isolation chamber inside a HALT chamber

removed from the chamber, a temporary thermal isolation box can be placed in the chamber enclosing the limiting subsystem. Temperature conditioned air can be provided to 'the chamber in the chamber' if needed, to keep the temperature within the needed operational temperature, as shown in Figure 6.9.

6.10.1 Other Configuration Considerations for HALT

In planning HALT, thermally limiting materials or technologies may have lower thermal FLT than active components in a circuit assembly. Some examples are LCDs, batteries and low temperature plastics which will soften or melt below 100°C and may limit the capability of the PWBA operation under high stress conditions. It can be useful to perform HALT with the limiting systems up to their UOL and LOL to observe any reliability issues within the technology limits before placing them outside the HALT environment and using extension cables to connect them to the UUT.

Covers or chassis housing circuits should be removed for thermal HALT. The high velocity airflow in HALT chambers far exceeds that of typical chassis cooling fans and a chassis will slow the thermal transition rates on the circuit elements. Thermal HALT does not require mechanical attachment of the UUT to the vibration table. Thermal HALT may be applied to a subsystem or system by simply setting the UUT on wood blocks or other support off the table to allow air circulation around the UUT for faster thermal transitions.

It is important to realize that any time that vibration is applied during HALT the structural mechanics of the system assembly become a significant contributor to all the resonant frequencies that result in creating fatigue damage in the UUT. If the chassis is integral to the mechanical strength it should be on during vibration HALT to reproduce the same vibration resonance responses that would occur in the final assembly of the system.

For the same reasons circuit boards in HALT vibration should have similar boundary conditions and attachment points as it will be in a chassis or housing if possible.

6.11 Lessons Learned from HALT

It is important in using any test that there is feedback to the designers about the problems or reliability risk found and the corrective action that eliminated the problem. The lessons learned will help prevent repeating a design issue in the next product design. After performing many HALTs on a variety of systems the HALT user will begin to observe the different strengths and weaknesses of electronics circuit layouts and configurations. HALT should not be finding the same weaknesses in subsequent designs.

Charlie Felkins performed HALT for years on data storage systems at the Storage Technology Corporation. An example of lessons learned from HALT vibration regarding the placement of electronics components after many vibration HALT evaluations is shown in Figure 6.10.

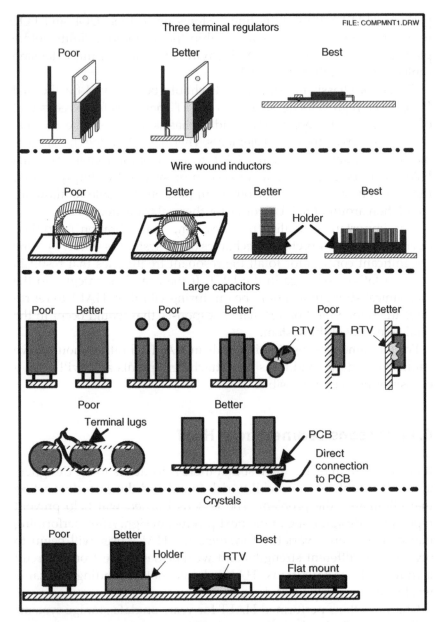

Figure 6.10 Good, better, and best component placements learned from HALT

6.12 Failure Analysis after HALT

If the product that is subjected to HALT is found to have limits at the FLT, then no investigation to find a cause of the limits is required. When a low margin or limit is discovered in HALT, or a limit with wide deviations between tested samples is found, the cause of the limit should be determined to know what will be required to increase the limit or to determine the reason for the wide deviation in limits between samples.

If a destruct limit has been reached in HALT on any sample the determination of the limiting circuit or component is easier, as the component destroyed is most likely the weak link. If the FA determines the component is the cause of the destruct limit, it should be replaced with a higher voltage or power rated component and HALT reapplied to see if the limit margins were increased.

Isolation of the cause or device that limits operational margins under stress can be a challenging process. It is most helpful to have the design engineer or other engineers who are very familiar with the circuit found to be the cause of a low limit. The easiest technique to help determine the cause of a thermal operational limit can be using an air hose from outside the chamber to change the temperature of a suspected component as being the cause of a low limit. A water cooled heatsink can provide local cooling and a high wattage resistor and power supply can provide local heating of a suspected limiting component to change its temperature after the LOL or UOL is found. Another device that can provide localized thermal control is a thermoelectric cooler – also known as a Peltier cooler – and power supply operate it.

If reaching the product's LOL and heating the suspected limiting component restores the system to operation the component being heated is the likely limiting element. Conversely, if cooling a suspected limiting component restores operation after reaching the UOL then it is the likely cause. It may be necessary to power cycle and reset the system after an LOL or UOL is reached and the suspected limiting component is cooled or heated above and below UOL and LOL to restore function.

Vibration failures that result in hard failures can be troubleshot on a bench and the failure point determined. Intermittent or soft failures that disappear when vibration stress is removed are more challenging to isolate. Tapping the circuit board and connectors may expose the failure point; also using a handheld vibration tool such as an electric metal engraving tool with a blunt tip can provide continuous localized vibration to stimulate areas of the circuit board while it is being operated.

Samples that have been through all the HALT processes should never be shipped to customers. It can be useful to keep the units used for HALT for future reference and comparison of design changes or reliability issues that may be discovered during the production life cycle.

7

Highly Accelerated Stress Screening (HASS) and Audits (HASA)

7.1 The Use of Stress Screening on Electronics

ESS processes have been developed since back in the 1970s as a way of eliminating latent defect failures, especially those defect mechanisms that cause failure in the first days or months of use. Due to the fact that temperature was considered the dominant stress that caused, failures of electronics, the first screens consisted operating a product for many hours at some steady state temperature above expected end use temperatures. Because of the steady state higher temperature operation, this stress screening came to be called 'burn-in.'

Any environmental stress to a new electronic component or system will precipitate some failures. The 'burn-in' of running a system for hours at a temperature higher than the use conditions would accelerate mostly latent defects that are caused by chemical reactions, as would be the case with oxidation on an connector causing high resistance. Burn-in at a constant temperature is not very effective at accelerating the detection of "cold" or cracked solder joints, delamination of

Next Generation HALT and HASS: Robust Design of Electronics and Systems, First Edition.
Kirk A. Gray and John J. Paschkewitz.
© 2016 John Wiley & Sons, Ltd. Published 2016 by John Wiley & Sons, Ltd.

component encapsulation or defects in material bonds. Thermal cycling is a more effective stress for the acceleration and detection of latent defects in material bonds through cyclic expansion and contraction of differing material interfaces. In a typical circuit board assembly, such as a PC motherboard, the number of material bonds and the potential for latent defects in the many material interfaces is much greater than those caused by chemical reactions.

In electronics product development, designing a robust product using HALT is very cost effective in the overall goal of making a highly reliable product design. The more capability you design into the product by removing low margins and getting as close to the FLT, the more tolerance in the design to withstand manufacturing strength variations in components and subsystems. It also results in the larger tolerance of end use stress variations of temperature, voltage and shock and vibration in large fielded populations. For reliability development, HALT has a much higher ROI over HASS, and HALT should always be used during new product development to make a robust design even if HASS will not be performed in production testing.

In the design of a system there are many design tools such as FMEA (failure mode effects analysis), FTA (fault tree analysis) and design reviews to help catch and correct design weaknesses and errors before hardware becomes available. Testing is the most beneficial tool when the first hardware becomes available which provides the most useful reliability and capability data. HALT is a very efficient and effective system test because it is a comprehensive test of the electronic and mechanical strength of a system and discovers exceptions to the high strength and capability and entitlement of correctly built electronic systems.

After the product has completed the HALT process and the development of margins to the FLT is as close as possible, the margin improvement process ends and the product is ready for production. The decision to perform HASS will involve many factors including the cost of failures for the customers, the maturity of the manufacturing processes and other important considerations based on consumer expectations. HASS is for precipitating and detecting latent defects that occur due to manufacturing process and therefore will be most valuable during the ramp-up in production volumes, where there is higher probability of errors being made during a new product manufacturing learning curve.

7.2 'Infant Mortality' Failures are Reliability Issues

The vast majority of latent manufacturing defects manifest themselves during what is commonly called the infant mortality region of the life cycle bathtub curve. The term infant mortality implies that the causes of failures are intrinsic to all electronic systems within the first weeks or months of a product use. Unfortunately, the term is mostly used dismissively as if the failures that occur during the early use period were expected.

The period of infant mortality is typically days or months, but for the most part it is defined arbitrarily by the manufacturer. The term is a human life terminology but the reasons for failures of electronics during the early use period have a different basis. Unlike a human after birth, an electronic component or system is not weaker when fabricated, becoming stronger with time; instead it has the highest inherent strength when turned on for the first time. Opposite of humans, electronics are 'adult' when first produced, and decline in strength (fatigue life) from that point on.

Many traditional reliability engineers pass blame for the infant mortality failures as a quality department problem, not to be confused with reliability engineering. The end user cares little about which company department is to blame for failure; the impact to the user is the same.

Latent defect mechanisms from manufacturing are undesired and therefore have an uncontrolled distribution in strength. The weakest manifestation of an infant mortality mechanism causes early failures within days or months and a strong manifestation of the same infant mortality latent defect mechanism may take a year or more to become a field failure and some latent defects are still strong enough never cause failure before the system is replaced due to technological obsolescence.

Early life failures in the electronics can be the most significant contributor to the overall life-cycle costs of failures for a company, especially if it is during a new product launch. Early life failures, occurring within the warranty period, result in measurable costs that include service calls, shipping parts and the raw costs of materials.

A much larger potential cost to a company during new product introduction (NPI) is loss of market share due to a reputation for poor quality. The internet has amplified the impact of poor reliability during the product's introduction to the market. With the internet and customer feedback being so easily accessed, unhappy customers' reports of poor reliability can be rapidly and widely dispersed around the world. The impact of internet customer feedback about poor quality on lost sales is difficult to quantify, but losing market share results in greater monetary losses over a longer period of time than the warranty service and material costs.

The only difference between HASS and HASA is in the percentage of production units it is applied to. HASS is applied to 100% of the products to be shipped, whereas HASA is performed on a portion of products to be shipped. To simplify the terms used, whenever the term HASS is used in this book it will also imply the HASA process unless explicitly stated.

HASS is developed from the operational and destruct limit data derived in the last HALT iteration. HASS is developed by using as many combinations of stress including rapid thermal cycling, random multi-axis RS (repetitive shock) vibration, power cycling, load cycling, and any other stress that makes sense and will help detect changes in the manufacturing processes that may result in field failures.

Some latent defects – such as a fracture across a BGA solder joint as was shown in Figure 6.4 – may only be found under stress conditions that cause an open circuit condition, and are undetectable when stress is removed, so they are only revealed by continuous monitoring in HASS. A very important factor in using HALT and HASS is that the UUT is operated and monitored throughout the HASS process.

7.2.1 HASS is a Production Insurance Process

HALT is the process of creating robust margins, and HASS is the process of continually ensuring that the robust margins are present in the shipping population during the manufacturing processes. HALT and HASS are not simulations of expected field environmental conditions, but are stimulations that use a small amount of the product's total fatigue life to expose and detect latent defects that comprise the front end of the life cycle bathtub curve. HALT is a step stress evaluation to discover the boundaries or limits of operation, and HASS is a process of rapidly applying stress, consuming a small amount of fatigue damage, and verifying

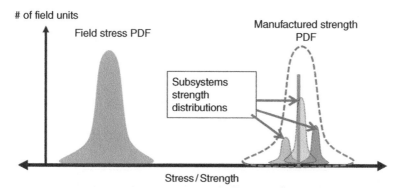

Figure 7.1 Stress/strength diagram with subsystems strength distributions

that operational margins have not declined due to ongoing changes that may occur in manufacturing or from engineering changes to the design. Any significant engineering design change needs to be evaluated by performing a re-HALT of the product.

The introduction of a latent defect and its impact on reliability can be shown using the stress/strength diagram. The strength PDF (probability density function) is a sum of the PDFs of all of the components and subsystems that are included in Figure 7.1. A decrease in the stress/strength margin can occur at any time during the production phase with a change in component suppliers, in the production line or facility or in manufacturing methods.

Any of the distributions of the components or subsystems' strength or capability declines because its distribution curve shifts left (becomes weaker) the system has a higher probability of failure as shown in Figure 7.2.

A diagram of the electronics life cycle bathtub curve (Figure 7.3) can help to illustrate the use of HASS.

The decision to use or not use a HASS process must be made based on many considerations such as:

1. the costs of the product
2. the costs of product failure to the customer in the field
3. the maturity of the manufacturing process and technologies
4. the ongoing costs of HASS.

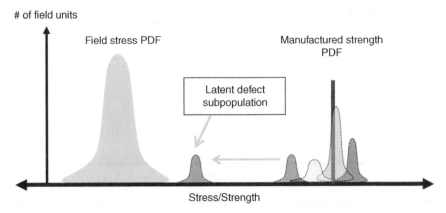

Figure 7.2 Stress strength diagram with subsystems latent defect distributions

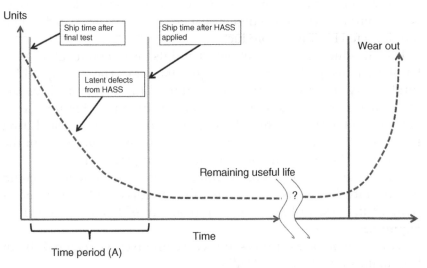

Figure 7.3 HASS uses some fatigue life to precipitate latent defects

If it is decided that HASS should be used for a new product, then the goal should be to make it as cost effective as possible. The higher the applied stress to the product, the faster a latent defect will be driven to a patent defect or easily observable failure.

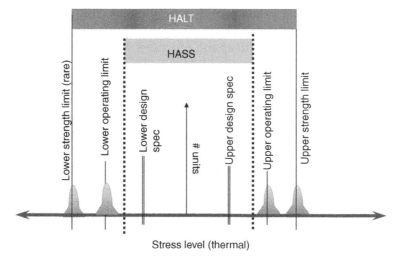

Figure 7.4 Stress levels for HALT and HASS

To be the most cost effective a HASS process needs to use the highest possible stress levels and the most combinations of stress that can be applied for the shortest effective duration. Completing HALT provides the product's stress operational limits and also the stress destruct limit which will define what safe stress levels can be applied in HASS.

An illustration of stress levels for HALT and HASS is shown in Figure 7.4. The normal curve of the distributions of limits represents the variation in strength across production units dependent on manufacturing process capability. The stress level used for HASS is set to be less than the potential distributions of operational limits so as to prevent a unit within normal variations being declared a failure in HASS.

7.3 Developing a HASS

After the last iteration of HALT and after the product development is complete a HASS process can be developed.

The steps for HASS process development are as follows:

1. Complete HALT after all the changes to increase margins have been made to the product if limits are not at or near the FLT.

2. Document the last thermal UOL and LOL and vibration UOL and UDL for defining the safe HASS stress regime.
3. Determine what test routines and functions will be exercised during the application of stress during HASS. The amount of functional coverage during the HASS process is critical as many defects and weaknesses will only be detectable under stress conditions and only if the circuit's function is being monitored. Trade-offs will sometimes be necessary between the amount of coverage, the duration of the HASS process and how easy or difficult it is to have the necessary auxiliary test systems near the HASS chamber.
4. The temperature cycle maximum HASS stresses are always lower than the thermal UOL or UDL. The derivation of the amount attenuation of the stress below the operational or destruct level has been made on the imprecise rule of thumb from many HASS users over many years. More specific derivations might be developed if the distribution of UOL and UDL levels are known from a set of HALT samples larger than three to five. Typically the maximum thermal level is 10–15°C below the UOL and 10–15°C above the LOL. The maximum input level of vibration is, again from a long used rule of thumb for HALT to HASS, limited to one half the vibration UDL in units of GRMS. A graph of the comparison of HALT and HASS stress levels is shown in Figure 7.4.
5. The HASS process is typically three to five thermal cycles along with other stresses run during the thermal cycles. A simple HASS stress regime showing precipitation and detection segments is shown in Figure 7.5.

7.3.1 Precipitation and Detection Screens

In some electronic systems there may be devices that have inherent limits to the technology that prevent its operation under high levels of stress or that the performance or output is degraded or skewed, so even though it partially operates we define that level of stress as the UOL. Examples are an analog audio amplifier circuit that has a continuous decline in gain when thermally stressed at high temperature. In digital systems it may be that a bit error rate exceeds a certain threshold that prevents stable operation. Operational failure in an audio amplifier may be defined as the temperature that the output

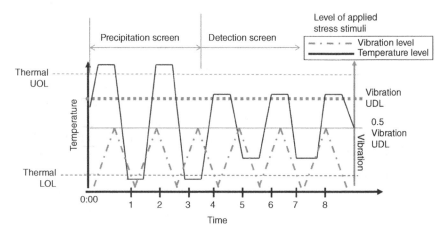

Figure 7.5 HASS precipitation and detection screens

drops to below 80% of the output level that it would have at room temperature; this is then the UOL.

Some technologies such as a spinning computer hard disk cannot operate during the application of random vibration, yet can withstand significant levels of vibration when the unit is powered off. Most LCDs used in consumer electronics have inherent thermal operational limits at about 70°C, which is well below what most its operating circuits are capable of.

Gregg Hobbs advocated using a two-level HASS process in the cases in which there is a significant difference between the HALT operational and destruct limits. During the first part of a two part HASS regimen, the precipitation phase, the levels of stress are above the UOL but below the UDL. The 'precipitation' phase of the HASS increases the rate of stress to more quickly precipitate a latent failure mechanism to a detectable state in later stress cycles. In the next part of the HASS, the thermal cycling and vibration stress levels are lowered to the just below operational limits. The system is then functionally tested in the 'detection' phase for any failures while still under albeit lower, stress conditions than the precipitation phase. An example of a combined precipitation and detection screen in HASS is shown in Figure 7.5.

The main goal in HALT should be focused on improving operational stress limits that will allow for HASS stress levels to be close to the fundamental limits of operational technology. Running a precipitation screening level above operational limits, where the product's functioning cannot be confirmed, should be avoided if possible. When a HASS is performed on a product at stress levels beyond the ability to monitor the function of the UUT, there is a risk of precipitating a defect but not detecting it, and then shipping the latent defect to potentially be a field failure. If the operational limit under thermal stress is limited by a designed-in protection circuit such as an OTP, then the OTP should be disabled for HASS if possible to allow for the highest thermal stress temperature cycles.

Vibration characteristics created with pneumatic RS hammers and the shift of the fundamental repetition frequency and the harmonic peaks seen in a typical PSD plot of the vibration table may have contributed to the observation that some defects, especially those stimulated by vibration, can only be detected when the peak of a harmonic vibration stimulates the natural resonance of a defect site. The shift of harmonic peaks of energy as the vibration intensity is increased in a RS system is explained in more detail in section 7.4 of this chapter.

HASS is used for high value, lower volume products and HASA for high volume, lower value products. A HASS process may be used for a relatively low cost subsystem if a subsystem failure can result in costly failures of a large system.

The learning curve for the manufacturing of a new system needs to be steep to prevent failures early in a new product's introduction to the market. Therefore HASS – if it is determined that it needs to be performed – has the highest benefit early in the production phase when the production volumes are ramping up. It is during this early phase of production that the probability of process problems are the highest. Any new manufacturing technologies introduced during this time will increase the probability of manufacturing-caused reliability issues.

The combination stress levels and durations are the same for HASS. The only difference between the two terms is the quantity of production volume that is subjected to it. HASS is applied to 100% of the manufacturing volume while HASA is applied to a sample percentage of the production lots. HASS is used when the quality levels are not meeting the desired goals. HASA is used when ongoing manufacturing quality levels have been achieved and the manufacturing process is stable. Switching from HASS to HASA should only be done when the yields through HASS have increased for known reasons, otherwise the causes of latent defects may return.

HASS is an active inspection process, but as with most inspection processes it is not capable of detecting 100% of the latent defect mechanisms. Generally the weakest manifestations of any latent defect mechanism is detected in a HASS and, when detected, initiates FRACAS (failure reporting and corrective action system) which should result in removing the cause of the latent defects. The probability of detecting a manufacturing subpopulation of products with latent defects is dependent on the quantity screened. Developing sampling rates for HASA to statistically detect a specific percentage of defects in a population can be created and followed, but this increases the chance of a manufacturing defect escaping detection. If a small but significant number of units with latent defects bypass the HASA process and later fail in use with the most important customers, it will be very difficult to defend to upper management the use of HASA to prevent latent defects from being shipped to customers.

HASS is a production test and, unlike HALT, is a pass/fail test. HALT samples should never be shipped to customers. In contrast, all units that pass HASS are to be shipped to customers as new product.

HASS on the other hand is essentially an ongoing manufacturing reliability insurance process. The decision to use HASS and when to end a HASS process should be based on the maturity and stability of the manufacturing process, the cost of the product and the costs of failure for the customers. The costs of a HASS process may not be justified for small consumer based products, but it still may have a significant benefit if HASS is applied early during the production ramp-up to full volume production. It is especially beneficial if the product includes any new technologies, new assembly methods or a new manufacturing facility in which there will be a learning curve period, which has a higher risk of manufacturing errors.

It is not possible for HASS to either precipitate and detect latent defects in 100% of products passing through the screen which is true of almost all final inspections or tests. There are several reasons for this. The creation of latent defects is undesired and uncontrolled. Therefore the process that creates the latent defect has a distribution of strength of the latent defects produced. Some percentage of the strongest latent defects will be strong enough to survive the HASS process without being precipitated and detected. Other reasons can be that the operational test applied during the HASS may not access 100% of the functions of the circuit, or a latent defect may only be detectable when a particular combination of stress conditions such as a certain temperature and voltage, or temperature and vibration, make the defect detectable.

The risk–benefit of HASA can be difficult to quantify, and many companies prefer to err on the side of keeping a HASA process in place and accepting the expense offset by the protection from a detectable major latent defect that could cost many times the expense of a HASA process.

The decisions to use HASS as in all engineering efforts require careful assessment and analysis of the risks of a latent defect being introduced to the manufacturing process, and that is unique for each type of system.

7.3.2 Stresses Applied in HASS

During HASS as many stresses as practical should be applied simultaneously during the process.

Typical stresses used in HASS are:

1. temperature cycling
2. vibration (modulated if using a RS vibration system)
3. power cycling
4. load cycling
5. clock frequency margining
6. other stresses unique for the product (pressure, motor speed variations).

HASS should not be needed for subsystems that have demonstrated field reliability and have not been changed, and the interface to the larger system is also proven to be reliable in the field. HALT may still be run to confirm the stress margins found in the first HALT of the subsystem, and confirm robust interoperation with the new higher-level assembly.

When the production is stable and capable, little benefit with HASS processes may be observed. Since most manufacturers are seeking to continuously lower the costs of manufacturing, there will be economic reasons for reducing or eliminating HASS. It is only when they discover a latent defect being introduced during manufacturing that a HASS process returns value. The causes of failures in HASS should only be due to manufacturing variations, or assignable causes, and not design weaknesses which should be found during HALT.

Before a HASS process is applied to products, it must be demonstrated that the screens do not consume a significant amount of the product's fatigue life. The verification process to demonstrate the safety of the HASS screen to be used for shippable product is called the safety of screen (SOS) process.

7.3.3 Verification of HASS Safety for Defect Free Products

A major concern for any stress screen process is being assured that the screening process does not consume a significant amount of fatigue life that would result in wear out failures during use. Remove too much fatigue life from a product carries the risk of its premature failure during the expected use period. HASS processes are developed from empirical limits found in a HALT procedure. Unlike ESS – which was developed from stress screening strength curves and the expected number of latent defects based on historic failure rates based from industry consensus – a HASS process is developed uniquely based on using the product's empirical strength capability. Using the full operational capability strength allows the HASS processes as stressful as can be achieved safely and to make the HASS process time as short as possible, which results in the most cost-effective screens. HALT provides empirical data on the maximum levels of stress for the strongest and shortest duration HASS.

The SOS process, just like HALT and HASS processes, is based on empirical test results. SOS is simply applying the proposed HASS stress regime to new units with multiple repetitions, typically 20 to 50 applications of the HASS regimen that will be applied in production testing, and verifying that significant fatigue damage has not occurred through careful visual inspection and thorough operational testing.

The underlying principle of the SOS is that if its total fatigue life was used with the typical SOS application of 20 repetitions of the HASS stress regime, then one application of the HASS regime represents using only 1/20 of the total fatigue life, leaving 95% of its fatigue life for the user. In practice, an SOS of 20 repetitions of HASS does not consume all the fatigue life. Equation 7.1 is the expression for determining remaining life after X consecutive applications of a single HASS stress regime to the new product samples:

$$Life_{\text{remaining}} = Life_{\text{total}} \left(1 - \frac{1}{X}\right) \tag{7.1}$$

Again HALT or HASS or the SOS cannot provide a time to failure or remaining useful life (RUL). It is because, the acceleration of fatigue mechanisms across a densely populated PWBA are accelerated at different rates under combined stress conditions and the probability density function of the distribution of an electronics product life cycle environmental profile (LCEP), it is very difficult if not impossible to determine a time to failure or failure rate at a system level.

The number of repeated applications of a HASS regimen, from 20 to sometimes as high as 100 cycles, for the SOS is based on the manufacturer's comfort level and the need for assurance that the HASS process leaves a significant life remaining to ensure it will not fail during its end use after it is shipped to the customer.

7.3.4 Applying the SOS to Validate the HASS Process

The final iteration of HALT will provide the stress operational and destruct limits for thermal, vibration, power cycling, voltage or load margins, plus all other variable operational stresses to be used for the HASS stress regime. A typical number of thermal cycles used for HASS is 3–5, and a typical time duration is 1–2 hours, but this is very dependent on the product and application. If the operational stress limits are increased to the FLT, simultaneous precipitation and detection throughout the HASS regime can be an effective screen. A simultaneous precipitation and detection screen is shown in Figure 7.6.

The largest difference in stress conditions generated in a HALT chamber is the random vibration frequency spectrum. Random vibration in pneumatic repetitive shock multi-axis HALT chambers induces rapid cumulative fatigue damage. Variations in the spectrum can be significant from one vibration table in a HALT chamber to the next for the same manufacturer and even more so between different manufacturers of HALT chambers. The frequency spectrum seen in the graph

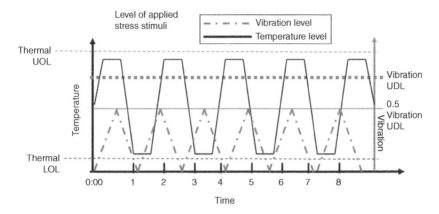

Figure 7.6 Stress regime of a typical HASS process

of PSD cannot be easily changed in multi-axis pneumatic RS vibration systems. The limitations of control of the frequency spectrum output of the RS vibration must be accepted, and each RS HALT vibration table has its own resonance peaks of energy across it's mounting surface, so it is possible to have significant differences in rates of fatigue damage generated across the table, for different design revisions from the same manufacturer, and even more difference between RS vibration systems manufacturers of HALT RS vibration systems. As a safeguard and to ensure that the vibration UOL in the chamber used for HASS is close to the vibration UOL in the chamber where the HALT was performed, a vibration HALT should be repeated in the chamber that will be used for HASS. Another way to state this is that the vibration levels for HASS should be established from the vibration HALT limits on the vibration table that will be used for production HASS.

To determine the number of applications that will be used for the SOS, it will come down to your test team to determine the comfort level and demonstrated assurance that one application of HASS does not consume a significant percentage of fatigue life. There is no HASS-to-field-time conversion for the reasons discussed in previous chapters. Therefore the only method to determine the total amount of fatigue life relative to the amount of fatigue life that a single HASS application removes would be to apply the HASS regime until defect free units failed, due to using all the fatigue life of the system. The difficulty with running an SOS until good units begin to fail is that it may take a significant time and still would not determine what field equivalent time under stress application of a HASS regimen represents.

To apply the SOS on new samples of the product. make sure that the SOS is applied using all the locations on the vibration table in the chamber used for HASS that will be used for the HASS process. There will be also be some variation of temperature in each sample position used in the chamber. Each of the individual samples should have the temperature and vibration levels measured with thermocouples and accelerometers at the same location on or in the product samples to observe the differences in the product responses to the stress inputs before starting a SOS. All locations that will be used in the HASS system during the production HASS should have similar stress responses. There will be some variations between samples, but this typically is not a problem as it is not known what stress levels are needed for the many different possible latent defects created during production. Closer uniformity of stress between samples may be achieved by changing UUT orientation, locations in the chamber, or redirection of air flow.

To verify all units' function during and after the SOS, the SOS units should be subjected to and pass normal production final tests. Additionally, the SOS samples should be operated for a period of 24 hours or more to again provide assurance that no weak latent defects have been introduced by the SOS. After completing all final tests and operational verification the samples should have a thorough visual inspection to observe any signs of latent damage. Examples of latent damage may be a discoloration of the circuit board due to the high temperature of a particular component, or melting of insulation on a wire or connector. These types of damage should not occur if the HASS was derived from a properly performed HALT, but it is just another verification that materials and components have not changed since the last HALT. Since the units completing HASS are shipped as new, there should not be any cosmetic damage to the SOS samples either. The SOS samples should be considered to have used up their fatigue life and therefore should not be shipped to customers. At least a few of the SOS samples should be stored for the future as they can be useful references to help determine a cause of a limit change or failure mechanism that is found in future production units.

7.3.5 HASS and Field Life

A common question for many new users of HASS is how much time in the field one HASS application equals. The only way to determine how much percentage of fatigue life is consumed with each application of HASS would be to perform a test of applying HASS until failure occurs in a defect free assembly. In general, the knowledge of a particular products fatigue rates and complex interactions of external and internal stress distributions that affect fatigue life are difficult to determine from a combined stress HASS profile. This is the fundamental reason that HASS results cannot be used to provide accurate quantified estimates of field life.

7.4 Unique Pneumatic Multi-axis RS Vibration Characteristics

Precipitation and detection screens may not be necessary for detecting most defects as most failure mechanisms will be detectable when the stress is increased beyond the initial stress level that the failure was first detected. It is possible and likely that concept of precipitation and detection screens may have been derived from many test facility observations that detection of defects occurred more frequently when RS vibration levels were changed during the HASS application.

Pneumatic multi-axis RS vibration systems are the type typically used in HALT chambers. The broadband vibration created from the low fundamental repetition rate is achieved by causing the table assembly to resonate at various frequencies. A typical RS repetition frequency range for the pneumatic hammers mounted on the underside of the vibration table is 30–50 Hz and the resonance of the table produces frequencies mainly in the range of 200 Hz to over 10 kHz. The hammer frequencies are not synchronized and that helps to smear the resonance peaks on the top of the vibration table.

To recall, fatigue damage in the UUT is only created when the UUT resonates at its natural frequencies. The purpose of vibration stress in HALT/HASS is to stimulate the bonds and structures of the UUT into its natural resonance frequencies which in turn rapidly create cumulative fatigue damage.

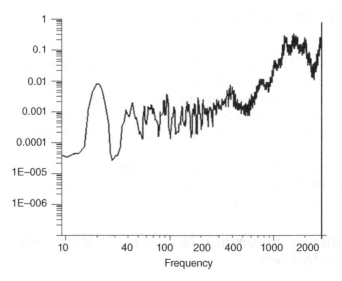

Figure 7.7 A PSD for the top of a multi-axis RS vibration table

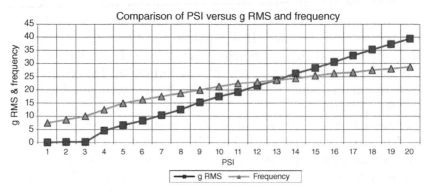

Figure 7.8 Correlation between air pressure and hammer frequency (Courtesy of Charles Felkins)

A typical RS vibration PSD is shown in Figure 7.7. The vibration energy peaks at the fundamental hammer repetition rate of about 20 Hz and at the resulting higher frequency harmonics. The vibration energy increases proportionally with the air pressure supplied as shown in Figure 7.8. Since the hammer fundamental frequency increases, the peaks and harmonics increase. The result is that the energy at peaks and valleys in the PSD will shift as the GRMS level increases frequencies. Manufacturers have reduced the 'picket fence' appearance of the RS vibration PSD of HALT systems and have smoothed out the peaks and valleys by making design changes to the vibration system, but some peaks and valleys are still present. Vibration resonance induces fatigue damage, which will stimulate a latent defect to become a patent defect. If the resonant frequency of the defect site is at the same frequency as one of the lower energy 'valleys' in the PSD, the defect may not be stimulated to a detectable failure. Modulating the vibration levels results in the peaks and valleys of energy in the applied vibration spectrum to shift in frequency and therefore have a higher capability of stimulating resonances at more potential defect sites.

During application of HASS the vibration levels should always be modulated throughout time RS vibration is applied. Modulating the vibration levels during HASS help to spread the vibration energy along the frequency spectrum. The resulting shift of the peaks and valleys of vibration PSD helps distribute the stimulation of more resonant frequencies of potential defect sites in the UUTs.

The correlation between the hammer frequency and vibration levels in g RMS is shown in Figure 7.8.

7.5 HALT and HASS Case History

Another experienced HALT user and reliability engineer, Mark Morelli, has been a strong advocate for HALT and HASS and has applied it to many different systems. Fortunately he has permitted the reprinting of a HALT and HASS a case history that he first presented in 2000 at NEPCOM West. The title of the presentation is 'Effectiveness of HALT and HASS' by Mark L. Morelli of the Otis Elevator Company.

7.5.1 Background

Highly accelerated life test (HALT) is a design 'ruggedization' test used to rapidly find defects in electronic products. Dr Gregg Hobbs of Hobbs Engineering Corporation [1] coined the term HALT. Highly accelerated stress screening (HASS), also coined by Dr Hobbs, is an ongoing test performed during production, used to find assembly and lot-related defects. More information on the techniques can be found in the references.

HALT and HASS were introduced at Otis Elevator Company [2] in response to a management challenge:

1. Do everything faster
2. Improve quality
3. Innovate
4. Reduce costs

This paper will provide examples of the reliability and cost benefits of performing both HALT during product development and HASS during manufacture. It is assumed that the reader has a basic knowledge of HALT, HASS or other methods of accelerated stress testing (AST).

HALT uses a 'test-to-failure' approach employing temperature, vibration, and electrical stress exposure [3] to rapidly precipitate and detect operating and destructive failures during product development. Those test failures deemed relevant, or likely to cause field failures, are eliminated from the product.

HASS is based upon HALT results and is a screening test used to find defects at the factory. HASS can be a 100% test but is more typically performed on an audit-basis, or highly accelerated stress audit (HASA).

7.5.2 HALT

At Otis Elevator Company, HALT was introduced in late 1995 and more than 100 tests have been performed since.

Before performing on new products, HALT was first performed on several products that had already been released, to correlate test results and actual use performance.

On one product line (an elevator motor controller), three of the top four and four of the twelve total types of field failures were observed during HALT. HALT found defects that comprised

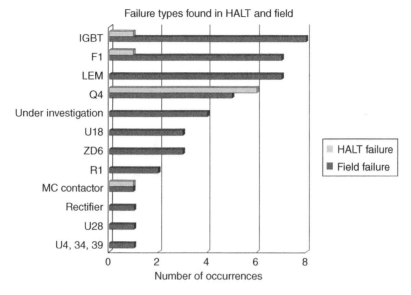

Figure 7.9 Otis field and HALT failures comparison

more than 50% of the total number of field failures. Analysis of the other failures indicated that other tests, such as EMI, humidity and HASS, needed to be performed to find the problem, or that the failure cause was eliminated on those samples tested. Figure 7.9 graphically depicts this data.

Based upon this initial testing, Otis continually monitors field performance on newly released products to verify that HALT is continuing to be effective. In addition, HASS was introduced because it was observed that not all failures could be found during development.

7.5.3 HASS (HASA)

HASS was implemented after field data analysis showed that not all failures could be found during product development. Some defects are introduced during manufacturing as the result of process variations or lot-related part failures.

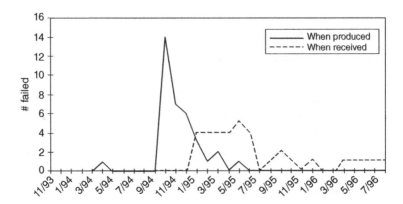

Figure 7.10 The effectiveness of a leaky capacitor introduced after HALT was completed

Figure 7.10 depicts a lot-related part defect (leaky ceramic capacitor). The reader can observe that most failures of this device occurred on products built during the months of November 1994 to March 1995.

The failures were not detected, and failed product were not returned to the factory for analysis until after February 1995, over a year after the design reliability tests were completed in January 1994. Since testing was not performed during manufacturing, there was no chance of finding this problem.

For the motor controller product line discussed in section 7.5.2, field failures have been eliminated (no field failures have been observed yet) on the samples (729) tested up to November 1999. The population of units not tested (5678) has experienced failures.

Figure 7.11 depicts the Weibull analysis [4] performed, which compares the reliability performance (cumulative failure percentage) of the tested and untested populations.

At the present rate of testing (average of 365 units per year), Table 7.1 shows the estimated total number of failures that will be prevented during a five-year production period:

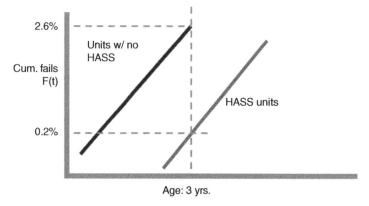

Figure 7.11 Comparisons of Weibull analysis of tested and untested populations.
Note: The cumulative failure rate of 2.6% at three years on the untested population correlates to the fallout during the two-hour HASS test in the factory

Table 7.1 Five-year estimate of failures prevented by HASS testing

Production Year #	Tested in HASS	#Failures prevented @age=3 years[1]	Cumulative # failures prevented
1	365	10	10
2[2]	365	10	20
3	365	10	30
4	365	10	40
5	365	10	50

[1] HASS fallout (2.6%) used to estimate # failures prevented; assumes no other reliability improvements except those identified via HASS process.
[2] HASS just completed on year # 2 production[5].

7.5.4 Cost avoidance

In addition to measuring the reliability improvement, Otis has also documented the cost avoidance that will be achieved as a result of implementing HALT and HASS.

Table 7.2 Cumulative cost avoidance of failures using HASS

Year #	Cumulative # failures prevented	Cost of failures (cumulative)			Cost avoidance[1]
		Visible	Non-visible	Total	cumulative
1	10	33k	83k	116k	87k
2[2]	20	66k	165k	231k	173k
3	30	99k	248k	347k	260k
4	40	132k	330k	462k	346k
5	50	165k	413k	578k	433k[3]

[1] Cost avoidance = Cost of failures − recurring cost of HASS. Recurring cost of HASS (not including chamber and support equipment) = \$80/unit × 365 units/year = \$29k/year
[2] HASS just completed on year # 2 of production
[3] Cost of HASS chamber & support equipment will be paid off after 4 years of testing!

The cost model incorporates three components of poor quality and the recurring costs such as labor and materials needed to implement HASS. The three components are:

1. visible costs
2. non-visible costs
3. hidden costs.

Visible costs are field labor charges for diagnosis and repair of failed equipment and the cost of building and shipping replacement hardware. Warranty costs are included in this category.

Non-visible costs include engineering and factory changes and continuing support required after a product is released.

Hidden costs include lost sales, loyalty and credibility with customers. Although difficult to estimate, hidden costs are certainly not zero. In one Otis region, hidden costs were estimated to be \$1M per year in the early 1990s!

For the motor controller product, Table 7.2 depicts the cost avoidance, or the cost of potential poor quality minus the recurring cost of HASS, achieved.

Bibliography

[1] http://www.hobbsengr.com
[2] 'Development of the Accelerated Stress Testing Process at Otis Elevator Company' R.V. Masotti and M. Morelli, Quality and Reliability International, Vol. 14, Issue No. 6
[3] 'History of Accelerated Reliability Testing at Otis Elevator Company, Part II,' M. Morelli, Proceedings of the 1998 IEEE Workshop on Accelerated Stress Testing
[4] 'Recipe for Reliability: HALT & HASS,' M.Morelli, Proceedings of the 1999 IEEE Workshop on Accelerated Stress Testing
[5] 'The New Weibull Handbook,' Dr Robert Abernathy

7.6 Benefits of HALT and HASS with Prognostics and Health Management (PHM)

The use of stress and measurements of a system under stress conditions for revealing hidden risks to human life is not new. Modern human health diagnosis and prognosis has made great advances from the use of imaging tools and technology for observing, recording and measurement of body chemistry and function. Tools and imaging help monitor human heath by monitoring physical changes to the body through measurement and analysis, which allows us to look for those changes that would indicate a health risk. The same approach can be used for electronics and mechanical systems for better prognosis of its health or risk of failure.

7.6.1 Stress Testing for Diagnosis and Prognosis

Stress testing is also used in human health to provide better detection of potential health risks, such as cardiac function. During a cardiac stress test, as seen in Figure 7.12, a patient will have their resting heart condition compared with the heart under high stress produced by intense exercise or through drugs. Measurements are made of the heart and body function while under stress, which may include blood pressure, blood flow measurement with an ultrasound echocardiogram, and electrocardiogram (ECG).

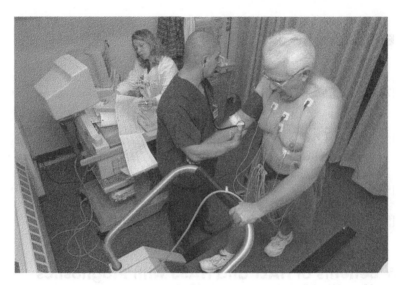

Figure 7.12 Cardiac stress test. Source: By BlueOctane at English Wikipedia (Transferred from en.wikipedia to Commons.) [Public domain], via Wikimedia Commons

In a similar approach, when electronic and electromechanical systems are forced to operational stress limits in a HALT, they will likely produce shifts in performance and critical parameters before the operational or destruct limits are reached. The parametric shifts should be similar with slight variations between samples under the same stress conditions. The range of the parametric shifts or stress induced operational characteristic curves discovered when using HALT can be applied in the field or in HASS as reliability discriminators, or indicators of potential failure in advance of actual catastrophic failures.

Accumulating and analyzing the shifts in performance or key parameters under stress conditions during HALT or HASS can be used for development of diagnostic and prognostic measurements. PHM methods can also provide the data necessary for modeling and determining the device or system's RUL.

A common prognostic that has been used with rotating machinery, such as motors, gears and bearings, is vibration analysis. For vibration analysis, an acoustical spectrum analyzer with accelerometer attached to rotating machinery can be used to observe the frequency characteristics of the vibration resulting during its operation. As the fatigue damage accumulates, resulting in the wear of bearings, loss of lubrication conditions or an introduction of foreign material (dust, dirt), the vibration spectrum of the monitored equipment will change. Analysis and comparison of the vibration spectrum or signature over time can produce leading indicator (diagnostics and prognosis) of impending failure before it occurs, allowing for preventative maintenance or replacement before becoming a catastrophic failure.

7.6.2 HALT, HASS and Relevance to PHM

In electronics, PHM is implemented by first discovering and determining a dominant mechanism(s) that will be a leading indicator or precursor to failure. HALT can be used to first find those weak links or dominant wear mechanisms that can be used for PHM.

The weak links found in HALT can be good candidates for finding the leading precursors to failure [1]. A common failure mechanism of BGAs (ball-grid arrays) is solder ball cracking. The cracked BGA solder balls will often lead to intermittent failures. It is very difficult to determine if a BGA has cracked solder joints causing failure without destructive failure analysis. Products having cracked BGAs in the field when returned to the factory may test OK on the bench after return. However, when they are monitored under thermal or vibration stress, the crack may be detectable as a high impedance or open circuit. A solder ball causing intermittent operation was demonstrated to be detectable under a combined stress HASS type regime by running self-test algorithms that measure a digital signal read and write. The test was also able to detect other failure modes such as wiring errors, loss of power to the board and cable damage [2]. These failure modes being detected while under HASS conditions is a good reason to use HASS for field returns that have a high rate of failures being reported yet cannot be reproduced when tested on the bench in ambient conditions. Showing how a HASS process can reveal an undiscovered intermittent failure mode that may have been found in returned products can be very useful for convincing management of the value of the use of HALT and HASS.

HASS is performed on new products to find future reliability failures. A key factor in the benefit of HALT and HASS is the monitoring a product's function during stress application. The measured

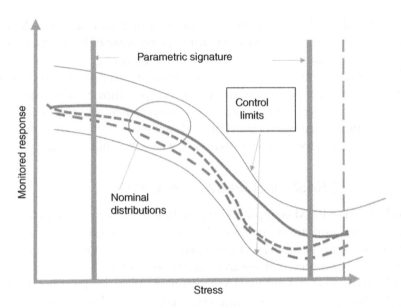

Figure 7.13 Generic parametric signature for reliability discriminators during HASS

parametric and performance variations between samples of the same product will be greater when under stress conditions than at ambient conditions. Therefore A HASS process has the potential to see variations in process control at a much higher resolution. Through comparing parametric signatures under stress conditions to an empirically determined nominal distribution, is a higher resolution reliability test that is more sensitive and comprehensive to shifts in product strength and risk of future failures at all levels of assembly before shipment. Continuously monitoring the product's function or key parameters during HASS processes and determining control limits, as used in SPC, will lead to better process control and initial reliability assessment from a operating characteristics (OC) curve as shown in Figure 7.13. A higher resolution of monitoring during HASS can provide a faster detection of process excursions.

Finding the relevant discriminators for operational reliability may be challenging. It may take many samples in HALT or HASS to determine the acceptable parametric signature control limits. Leading process indicators for reliability discriminators can be developed by parametric data collection of a product's operational response during HASS application. Shifts during stress application in voltage, the skew and jitter of a digital signal, signal propagation or power consumption could be variables that can be used as discriminators. Any product that has a stress response signature falling outside the acceptable limits could be a leading indicator of lost process control at some level of manufacturing. Samples that have intermittent or marginal operation issues may especially be useful for finding the relevant parametric signature control limits. The limits may be established based on the products' stress response to combination of input stresses (i.e. voltage, temperature, over or under clocking). PHM is a new approach to reliability monitoring and holds promise of making HASS processes more comprehensive and effective screens. PHM in HALT and HASS will require careful adaption to the wide variety of electronic systems, but as the principles of PHM and HALT and HASS become more widely used, systems may be included in the design, also known as design for testing (DFT), to have easier and more accessible monitoring of critical parametrics to see parametric responses to HALT and HASS stresses.

Bibliography

[1] The Useful Synergies Between Prognostics and HALT and HASS Silverman, M. and Hofmeister, J. Reno, Nevada: IEEE Reliability and Maintainability Symposium, 2012. 2012 Proceedings – Annual Reliability and Maintainability Symposium. pp. 226–230. 978-1-4577-1849-6.

[2] HALT Evaluation of SJ BIST Technology for Electronic Prognostics Hofmeister, J.P., Vohnout, S. and Mitchell, C. Orlando, FL: Autotestcon, 2010 IEEE, 2010. 978-1-4244-7960-3.

8

HALT Benefits for Software/Firmware Performance and Reliability

8.1 Software – Hardware Interactions and Operational Reliability

During the research for this book it became clear that published and available data and information on the interactions between hardware and software reliability are very scarce. It is not understood why this is so, as many engineers in the field of digital electronics are aware that some hardware problems can be fixed with software changes, such as a code change to delay particular code execution, and also some software failures can be fixed with hardware changes such as changing to a higher speed version of a particular active digital component or using a shorter cable.

Operational reliability can be affected by the quality of digital signal transmissions or signal integrity. The signal quality, timings, skew and jitter and electronic noise levels are affected by the variations in materials

Next Generation HALT and HASS: Robust Design of Electronics and Systems, First Edition.
Kirk A. Gray and John J. Paschkewitz.
© 2016 John Wiley & Sons, Ltd. Published 2016 by John Wiley & Sons, Ltd.

and manufacturing processes, along with environmental conditions such as temperature, voltage and humidity.

The vast majority of new electronic systems are digital. In the mass manufacturing phase of digital electronic systems there are many electrical parameters that will be affected by the stack-up of manufacturing process variations. Semiconductor device manufacturing is an imprecise operation. There are process variations in the many steps of fabrication of silicon die that will affect parametric performance from die to die on the same wafer. In deep submicron processes, variation of transistor threshold voltage can produce over 30% sheet resistance, and variation of poly-silicone resistors can reach 40% for some technologies [1].

In the manufacturing of CMOS semiconductors the main sources of variation come from gate oxide thickness, which may consist of single digit numbers of atoms thick, random doping fluctuations, device geometry from lithography in the nanometer region and transistor threshold voltage. The variations can be a range of 100% for threshold voltage across a chip, 30% speed variation across a wafer and 100% leakage current variation in a wafer manufactured with 130 nm transmission line widths [2].

There is a distribution of values within the minimum and maximum required electrical performance parameters for semiconductor devices. For some semiconductor devices the measured parametric deviations from process variations can be used for product 'binning'. In semiconductor devices product binning is a process where the measured thermal and frequency differences in a device's performance using the results of specific algorithms in categorizing of finished products for different markets.

The timing and quality of signal transmission is dependent on the stack-up of parametric variations starting at the level of the silicon die up to the final assembly of the whole system and interconnections to other systems.

Stimulating timing variations of active devices in a digital system can push marginal timings to reproducible operational failures. Thermal stresses in HALT and HASS stimulates shifts in the speeds of signal propagations. In thermal HALT, stress levels are increased with temperature steps down to the lowest and up to the highest empirical operational limit, which results in recoverable failure.

Operational reliability can only be improved by reproducing marginal failure modes so the cause can be determined and corrected. Shifting propagation signal speeds and timings of active devices in a digital system can push marginal timings to operational failures.

When a low thermal operating limit is found in HALT, applying individual heating and cooling to the suspected active digital component can help isolate the cause of the thermal limits, both hot and cold.

Manufacturing variations and aging in hardware can cause operational ('soft') failure or degraded performance, but may be elusive to reproduce or detect. The shrinking of feature sizes, lower voltages and higher clock speeds for digital systems will likely result in more software operational reliability sensitivity to the stack-up of hardware variations in fabrication of components and system assembly.

The same marginal timings found in thermal HALT could result in marginal operational reliability which is difficult to find when bench tested upon return. It is probable that many NFF or CND returns are a result of marginal signal integrity. Since soft errors do not require hardware replacement, the cause of them may never be known and they are therefore tolerated even though they may lower the potential performance of a system.

Much of the software development is done with dynamic software tests and static techniques such as code reviews and walkthroughs before hardware is available. When the system hardware becomes available the systematically conducted unit and system tests are used to reveal faults that lead to failures during software execution. It is difficult to discover an underlying cause of software failures if they are not easily reproduced. Software testing is best suited to finding bugs that consistently manifest themselves under well defined conditions and generally stop the execution of the software. In the software testers' jargon, these faults are referred to as 'Bohrbugs', an allusion to Niels Bohr's simple atomic model [3]. Because the Bohrbugs can be consistently reproduced by using the same sequence of operation, they are the easiest to determine the cause of and eliminate.

And then there are software bugs that cannot be easily reproduced. Using the exact same software test execution sequence sometimes reproduces the fault, but sometimes not. The software bugs that seem to disappear the moment that you start to look at them are called 'Heisenbugs' by the software testers in an allusion to the Heisenberg uncertainty principle in physics. Heisenbugs are difficult to reproduce and may elude bugcatchers for years of operation. In fact, the slight changes from bugcatchers may perturb the operation enough to cause the bug to not occur again [3].

In digital hardware debug efforts, Heisenbugs and the difficulty of reproducing them in order to fix them can significantly delay the market release of a new system and cause loss of market share. Heisenbugs are likely to occur in digital systems that have marginal signal quality or marginal timings that will cause intermittent failure. Using increasing and decreasing temperatures, voltage and clock frequencies can push marginal timing conditions to operational failures and can help in reproducing many of the Heisenbugs. The faster that Heisenbugs can be reproduced the faster they can be eliminated and the faster the product can be released to market.

8.1.1 Digital Signal Quality and Reliability

Performance and operational reliability due to variations in devices up to system fabrication at the circuits and system level have not been often addressed in industry literature. As semiconductor geometries continue to shrink, bus frequencies increase and power and voltage requirements in ICs decrease, so process variations and the resulting parametric variations will become a more critical factor in producing an electronic system with high operational reliability.

Signal integrity in digital electronics is an important quality for the reliable operation of digital systems. Electrical signals that represent the binary language of 1's and 0's are in reality analog signals that vary in amplitude, noise, distortion and loss. The quality of the electrical signals is affected by the materials and geometries of conductive traces and layers in a PWBA. As the densities of integrated circuits and PWBAs continue to increase, and voltages and power decreases in electronics, variations in fabrication of ICs and PWBAs will have a greater impact on the quality of electrical data signal integrity and therefore operational performance.

Along with manufacturing process variations, the quality of the signal can be degraded by the progression of fatigue damage and chemical reaction mechanisms that occur over time. The signal path impedance will change and impact the signal quality. Degradation and aging can lead to intermittent failures which may be difficult to reproduce and isolate at ambient conditions. Thermal or vibration stresses can shift or skew the marginality to make it a persistent and detectable failure before the fractures in solder joints occur or a conductive trace becomes an open circuit and a patent failure.

8.1.2 Temperature and Signal Propagation

Signal propagation can be shifted or skewed by changing the temperature. The high to low propagation delay versus temperature for in a Fairchild octal buffer is graphically shown in Figure 8.1.

The graph shows a single device measurement. If a large number samples are measured, we would expect to see a distribution of the variations of the propagation delay as shown in Figure 8.2. Variations in the fabrication process will lead to lot-to-lot variation in the propagation delay. How much future variation will occur, or how well centered within specifications during mass production is a future unknown. Circuit simulations can be used to determine circuit sensitivity to the worst case propagation values with ideal device models. The mathematical models used to simulate the thermal effects on the electronic device and system parametrics are approximations, and the real system will be affected by manufacturing and environmental variations.

Changing semiconductor temperature skews the signal propagation speeds in semiconductors and also conductors. The relationship shown

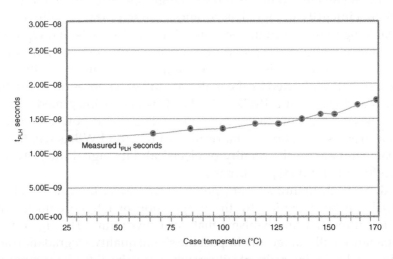

Figure 8.1 Measured low to high propagation delay versus case temperature of a Fairchild octal buffer. Source: Adapted from Condra et al, 2001 [4]

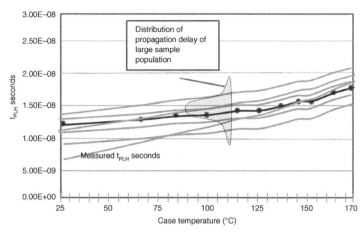

Figure 8.2 Potential distribution signal propagation delay in mass production. Source: Adapted from Condra et al, 2001 [4]

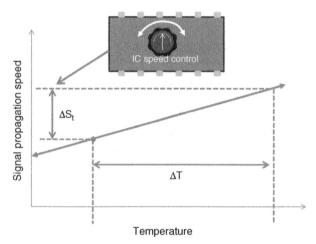

Figure 8.3 Applied temperatures skewing of timing distributions in semiconductors

in Figure 8.3 between signal propagation speed and temperature is another benefit of HALT, especially with thermal and voltage stimuli, helps to simulate potential speed distributions that will occur in the populations of mass-produced semiconductor devices and across signal transmission pathways used in the design and manufacturing of digital electronics systems.

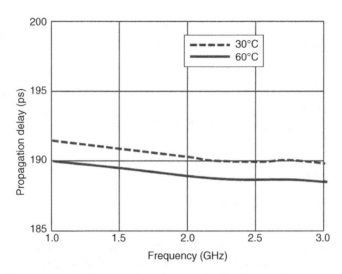

Figure 8.4 Propagation delays for short PCB trace. Source: Adapted from Yip et al, 2010 [5]

The effects of temperature on the speed of signals in electronics circuit board assemblies is presented in a paper titled 'Electrical-thermal co-design of high speed links' from Rambus, Inc. In the paper, the effects of temperature differences on the quality of data links are presented. Measuring the total jitter on a 3.2 Gps data link at a temperature of 0°C and then at 75°C showed an increase by as much as 10% of the nominal value. Figure 8.4 shows the differences in propagation delay of a short PCB trace.

Temperatures in real systems can and do have a significant distribution across the operating circuit hardware. The thermograph of an operating circuit board in Figure 8.5 illustrates that the temperature gradients across it can be greater than 40°C.

A different distribution of the temperatures across a circuit board in the identical systems can be significant if the system is located in a hotter or cooler environment. An example would be a desktop PC, which may sit in an open office area or be placed in a closed cabinet with little airflow. The temperature differences between the two user locations can be significant and will affect the quality of the signal. If the signal becomes marginal it may increase the BER (bit error rate) between elements of the circuit and will begin to impact functional speeds due to signal retransmissions and error correction algorithms.

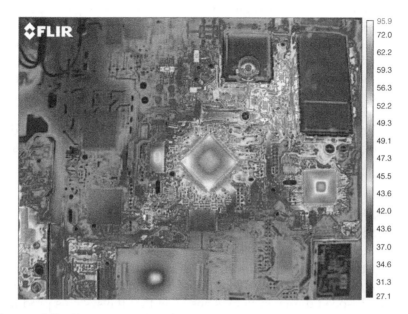

Figure 8.5 Thermograph of an operating circuit board showing thermal gradients across board. Source: FLIR Corporation, 2013. Reproduced with permission of FLIR corporation

8.1.3 Temperature Operational Limits and Destruct Limits in Digital Systems

The traditional thermal HALT protocol is to first find the LOL (lower operating limit) and then the UOL (upper operating limit), then the LDL (lower destruct limit), then UDL (upper destruct limit). After the limits of single stresses are found, thermal cycling and combinations of stresses should be applied to observe other potential weaknesses in the product.

Rapid thermal cycling is a good stress to observe marginal timing issues by temperature differentials forcing slight parametric variations between components.

Thermal HALT on digital systems almost always shows that empirical operational limit or failure is a 'soft' failure and rarely causes catastrophic damage to the hardware. This could be a common fear for engineers who have not witnessed thermal HALT on digital hardware to the point of failure.

Since thermal HALT on digital systems results in ceasing the operation of the system, any higher or lower temperatures will be applied to non-operational systems. Destruction of the system will occur most likely when the components or materials reach a change in material state as in plastics melting or solder reflow, which in almost all cases is a failure mode that is not a potential cause of failures in the field, and therefore an irrelevant failure mode in HALT. If thermal HALT operational limit results in a component failure in a digital system, understanding the cause of the destruct limit should be a priority as it is an especially rare weakness and therefore a higher reliability risk.

8.2 Stimulation of Systematic Parametric Variations

During the design of new systems, device models are used that are ideal in their performance. The challenge with the development of a new high speed digital system is ensuring that the systems produced will all have operational performance at the maximum processing speeds possible without failures.

Variation in manufacturing processes affects reliability and performance of digital systems. As the speed of digital hardware increases, materials and fabrication process variations will become a more significant contributor to a system's operational reliability. Controlling and limiting the variations through the components and up to system level manufacturing is of course the focus of statistical process control (SPC). SPC applied to components and assembly manufacturing is used to monitor the critical parameters and correct causes of variation to ensure that all devices manufactured will be well within the design's maximum and minimum specifications.

In complex electronic circuits there can be thousands of components and interconnections. Each level of assembly adds to the potentially complex parametric signal quality stack-up. All of the contributors to the variation will change over the manufacturing period of the product. Electronics design engineers realize this and account for the allowed variations from manufacturing through worst case tolerance analysis and CAD simulations of circuit performance through circuit simulation programs like SPICE. Yet the interactions between

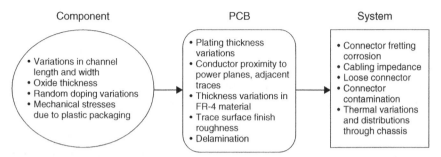

Figure 8.6 Potential contributors to poor signal integrity in electronics

variations in the real system may not always be considered or known in the mathematical models used for CAD simulations.

The potential manufacturing variations that can affect systems performance and reliability are shown in Figure 8.6.

The changes in parametrics may be only slight, and isolation of the circuit element that has the lowest margin affecting operational reliability can be difficult. Electrical probing of an operating subsystem apart from its main system may slightly shift or skew the environmental conditions, such as temperature, voltage or clock frequency which may restore normal operation of the system and mask a fault condition for the full system.

8.2.1 Parametric Failures of ICs

Parametric failures are some of the most difficult failures in CMOS ICs. They cannot be easily found in stuck-at, or delay fault or I_{DDQ} tests. Instead they are mostly speed related failures when they are used at different power supply voltages or temperatures from those they were tested at during manufactured. During the fabrication of CMOS ICs individual transistors and metallization processes result in variations within a die, die-to-die and lot-to-lot. The variations in fabrication can significantly change the speed of the circuits. Variations in channel length and width, gate oxide thickness, via and contact resistance all affect the circuit's signal propagation and therefore the overall operational speed. Metal interconnection variations that result in signal speed variations include metal width, thickness, spacing, granularity and current density [6].

Defects and statistical variations combine through the multiple manufacturing steps of system fabrication and can shift parameters from their design values. Dimensional variations in IC channel length and width and oxide thickness, along with random doping variations, directly modulate three very important electrical parameters: threshold voltage (V_t), transistor off-state leakage (I_{off}) and the drain saturation current (I_{Dsat}). Signal noise created from crosstalk and current switching also affect circuit speed, but can be prevented by following design rules and are generally not associated with manufacturing defects [6].

As bus frequencies of digital electronics increase, effects in data transmission that were second or third order in earlier designs begin to dominate as bus speeds continually increase in digital systems [7]. The variations in the multiple levels of fabrication that modify the signal quality and parametric performance of high speed digital electronic systems that are present in the actual hardware are complex, difficult to measure, may not be known and therefore are not accounted for in mathematical models. Every conductor has frequency dependent inductance and capacitance that impact to the quality of signal transmissions from each node of the non-ideal conductors.

Manufacturing and materials distributions in circuit board assemblies adds further variable modifiers to signal quality. A very simple representation of a cross-section of a typical FR4 circuit board in Figure 8.7 shows the copper transmission lines on the surface and in buried layers between the power and ground layers. Noise, crosstalk and reflections are dependent on PCB fabrication variations. As the signal speeds increase to gigabits per second the PCB substrate can no longer be assumed to be homogenous. The fiberweave effect of typical FR4 fiberglass makes the signal insertion loss dependent on the relative humidity. The result can be a dramatic difference in signal quality between a system being operated in dry Arizona or humid Malaysia.

Every copper trace has a frequency dependent inductance and capacitance that impacts the signal quality from each node of the non-ideal conductor. Each transmission line has magnetic and electrical fields that create noise and crosstalk in adjacent traces. The copper layer must have some surface roughness to provide adhesion to the

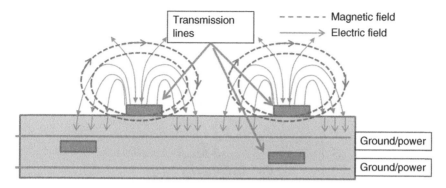

Figure 8.7 Cross-section of simple circuit board

FR4 layers. Surface roughness increases the conductor resistance and inductance, and trade-offs must be made to ensure high speed signal quality [5].

8.2.2 Stimulation of Systematic Parametric Variations

During new product development, initial prototypes and pilot product quantities are generally low. Manufacturing variations between prototypes will therefore be lower than the potential variations during high volume mass manufacturing. Mass manufacturing, especially for high volume systems such as consumer electronics or IT hardware, may be with a mix of component first, second and maybe third sources of materials and components.

The graphs in Figures 8.8 and 8.9 illustrate the distribution of a hypothetical timing parameter in a limited number of samples of the first hardware builds. The devices and systems built in the prototype or pilot phase will have minimal variations of signal propagation speeds relative to variations in mass manufacturing, as the assemblies are built with similar date codes and on the same assembly line.

As the volume of production increases the stack-up of the parametric timing distributions will increase as second and third sources of components are mixed into production, along with the variations of the manufacturing assembly processes such as the number of manufacturing lines increase.

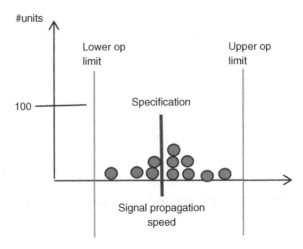

Figure 8.8 Potential distributions of parametric timing variations during prototype/pilot production builds

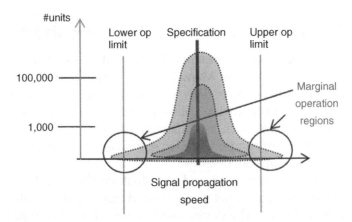

Figure 8.9 Mass production and the wider distributions causing marginal operation

Thermal stress shifts or skews the electrical parametrics throughout a system. Lowering the temperature of the prototype circuit elements skews the signal propagation speed to the higher end in a small sample population used for thermal HALT as illustrated in Figure 8.10.

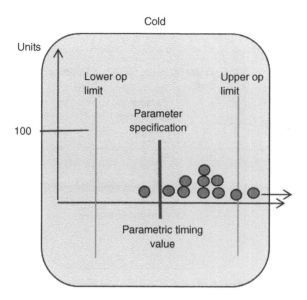

Figure 8.10 Colder temperatures skew signal speeds higher in a sample of prototype hardware

Raising the prototype's temperature skews the signal propagation speeds lower in the small sample population signal speeds as shown in Figure 8.11.

Thermal stress skews the impedance of all the signal conduction paths and the parametric functions of the circuit components. This functional skewing of circuit elements may help simulate the effects of lot-to-lot variations in circuit boards' and components' parametric functions before they cause operational failures in some portion of the shipped products. For this reason rapid thermal cycling should be used during HALT of digital circuits. Rapid thermal cycling applied in a HALT chamber creates thermal gradients spatially and temporally. Forcing temperature differences between components creates differential timing shifts between active components, skewing the timings and other parametrics across all circuit elements to discover marginal signal integrity failures.

Thermal cycling stress in HALT and HASS has the benefit of forcing rapid fatigue damage from the expansion and contraction of material bonds and interfaces. Thermal cycling adds another synergistic benefit

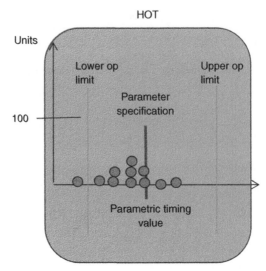

Figure 8.11   High temperatures skew signal speeds lower in a sample of prototype hardware

resulting from thermal expansions in material dimensions and dimensional variations, along with conductor impedance, and this skews the signal quality which can show additional weaknesses in operational reliability. Rapid thermal cycling creates thermal gradients across a circuit board or system, as shown in Figure 8.12, providing the comprehensive synergy of shifts in material dimensions and device parametric values.

During the mass manufacturing of the circuit there will be the uncorrelated distribution of all of the parametrics of active devices along with the contribution of variations in board fabrication and assembly. In time, fatigue damage to circuit board transmission lines and materials may also contribute to changing signal quality and therefore operational reliability.

Stimulating timing variations of active devices in a digital system can push marginal timings to reproducible operational failures. Thermal and voltage HALT methods can help improve operational reliability by discovering marginal digital circuits before wider variations from component suppliers result in operational unreliability in some percentage of manufactured and

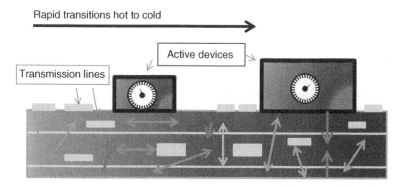

Figure 8.12 Rapid thermal gradients shift dimensions and parametrics of active devices

shipped systems. When a low thermal operating limit is found in HALT, applying individual heating and cooling to the suspected active digital component can help isolate the cause of the thermal limits, both hot and cold. An example of the benefits of thermal HALT helps to detect software operational is presented in the next chapter.

Bibliography

[1] Melikyan, V.Sh., Durgaryan, A.A, Balabanyan, A.H., Balayan, E.H., Stanojlovic, M. and Harutyunyan, A.G. Zlatibor: s.n., *Process-Voltage-Temperature Variation Detection and Cancellation Using On-Chip Phase-Locked Loop*. 2012. Proceeding 56th ETRAN Conference.

[2] Patel, J.H. Youtube Video: CMOS Process Variations: A Critical Operation Point Hypothesis. Youtube. [Online] September 26, 2008. [Cited: 31 May 2013.] http://www.youtube.com/watch?v=rf8qTpW6BH4.

[3] Gray, J. s.l.: *Why Do Computers Stop and What Can Be Done About It?* Proceedings from 5th Symposium on Reliability in Distributed Systems, 1986. pp. 3–12.

[4] Condra, L., Das, D., Pendse, N. and Pecht, M.G., Junction Temperature Considerations in Evaluation Electronic Parts for Use Outside the Manufacturers-Specified Temperature Ranges. 2001, *IEEE Transaction on Components and Packaging Technologies*, pp. 721–728.

[5] Yip, T.G., Beyene. W.T., Kollipara, G., Ng, W. and Feng, J. *Electrical-Thermal Co-design of High Speed Links*. Los Altos, California: IEEE, 2010. 2010 Electronic Components and Technology Conference. pp. 1893–1899. 978-1-4244-6412-8.

[6] Segura, J., Keshavar, A., Soden, J. and Hawkins, C. The Nature of Parametric Failures in CMOS ICs.

[7] Hall, S.H. and Howard, H.L. *Advanced Signal Integrity for High-Speed Designs*. s.l.: John Wiley and Sons, 2009.

9

Design Confirmation Test: Quantitative Accelerated Life Test (ALT)

9.1 Introduction to Accelerated Life Test

After the empirical limits of the design have been determined in HALT or similar AST, and weaknesses in the design have been corrected, there is often a need to estimate life in the application, or demonstrate the life of the product relative to particular stresses. This is especially applicable to suppliers of components or assemblies who may need to demonstrate that their product can meet the OEM customer's reliability needs.

Quantitative accelerated testing can be conducted at the following levels:

- materials
- components
- subsystems or subassemblies
- full system

Next Generation HALT and HASS: Robust Design of Electronics and Systems, First Edition.
Kirk A. Gray and John J. Paschkewitz.
© 2016 John Wiley & Sons, Ltd. Published 2016 by John Wiley & Sons, Ltd.

Usually testing at higher levels of system integration will result in less acceleration, and it is more costly and time consuming. So, most accelerated tests are run at lower levels of product integration, typically at materials, components or small assembly level.

In cases where the applicable stresses can be defined and simulated in a test and the UUT is subjected to one or very few stresses, this can be an effective way for suppliers to demonstrate the estimated life or reliability of their product. Caution is required as more complex products or assemblies subject to multiple stresses and levels are tested. Quantitative accelerated testing of complex systems with multiple stresses will require more samples and more levels of the stresses to perform analysis and extrapolate life or reliability estimates from the testing. This makes the testing more costly and time consuming. It is still more efficient than traditional reliability demonstration testing conducted at normal application levels. It also provides insight into the interaction of multiple stresses over repeated exposure which can precipitate degradation and wear-out failure mechanisms. If the combined stresses that the system is exposed to are difficult or expensive to simulate in the laboratory, testing in the customer's system and application (sometimes called beta testing) may be the best alternative for evaluating performance and life in the system. This also provides the opportunity to test interfaces between components and subsystems which are often the source of reliability issues in the field.

When appropriate for selected components or assemblies, life tests can be accelerated in two basic ways. The first is usage rate acceleration which is applicable to products that experience intermittent operation in the application and have significant idle time where the system is shut down. By running an operating profile or sequence repeatedly without the down time, the life exposure to stresses can be accelerated. The one caution for this type of accelerated test is that thermal cycles must be considered by allowing sufficient off time to cool the test units back to ambient conditions before starting the next profile cycle. These thermal cycles can precipitate degradation and failure modes in many materials and need to be included in the usage rate acceleration.

The second approach to accelerated life testing is increased levels of stress beyond nominal and extreme user levels. This is applicable to products with continuous operation or very little off or down time. In these cases, usage rate acceleration is not possible or sufficient to accelerate the test. So, accelerated stresses are needed to accelerate degradation and failures in a shorter test period. The first step is selecting which stress or stresses to use in the accelerated testing. These are selected based on understanding the application stresses on the product and the results of the earlier DRBFM analysis of potential failure modes and causes. Once the team has agreed which stresses to apply in the test, the second step is conducting HALT or AST to determine product limits. This determines the design margin of the product and the range of accelerated stresses between normal application level and the upper limit that can be used for quantitative accelerated life test. Using the results of the HALT or AST testing, which identified stresses causing failures, the stresses to be applied in quantitative accelerated life test can be selected.

There are two basic methods to select the levels of each stress used in ALT. The first is to select three stress levels between the upper limit and target application use level that are spaced to define the life–stress relationship. Fewer samples are tested at the highest stress level and more samples are tested at the lowest accelerated stress level in order to provide a better definition of the life–stress relationship and a better extrapolation of life at the application use level. Samples are allocated to each of the three accelerated stress levels in a ratio of 1:2:4 for seven samples and 4:7:9 for 20 samples from highest stress to lowest stress as examples. The life–stress plot is shown in Figure 9.1 and illustrates the three accelerated stress levels and the application stress level in a log-log plot. This helps visualize the relationship of stress level to life or reliability of the test units.

The second method to select the stress levels for the test is a sequential process starting at the upper limit determined in HALT or AST and then reducing the stress at the limit to a level at a selected percentage below the limit. Then two or three samples are tested to failure at that level. Stresses are then reduced another percentage and a few more samples are tested to failure at this second stress level. This enables determination of a preliminary life–stress relationship. Extending the life, the stress plot to the

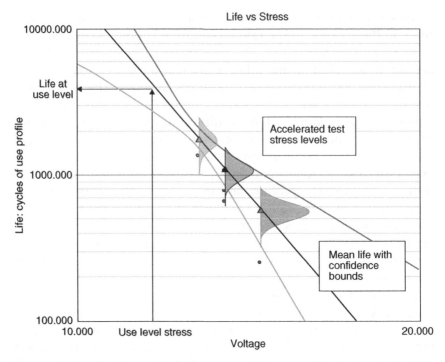

Figure 9.1 Life–stress relationship plot for quantitative accelerated life test

remaining time available for test determines the third stress level for testing. More samples are tested to failure at this level. So, the sequential method still allocates more samples to the lowest stress level and fewer at the intermediate and least at the highest stress. This enables a better extrapolation and better fit to the life–stress model used.

With either approach, analysis methods to determine the life–stress relationship and extrapolate life at the expected stress level in the application are the same. The process involves fitting the test data to a distribution, and selecting a life–stress relationship model that is applicable for the stresses applied. For example, this could be the inverse power law for mechanical stresses or voltage applied or it could be the Arrhenius model for temperature, or the Eyring model for stresses such as temperature or humidity. Other multiple combined stress models are also available for consideration. Refer to the works of Nelson, and Meeker and Escobar for details of the analysis methods [1,2].

9.2 Accelerated Degradation Testing

A variation of the overstress type of accelerated life is the accelerated degradation test. Overstresses are applied similar to the description in the previous section, but failures are defined as degradation of selected parameters below an acceptable threshold. Degradation testing with repeated measurement provides multiple data points and much more information than a single failure-time data point for a sample. This begins to address early wear-out failures. It also directly supports development of built-in or external sensors and diagnostics for condition based maintenance (CBM) or prognostics and health management (PHM). This enables the operator to be alerted to the need for preventive maintenance. It can be applied to electronic, electrical and mechanical products as long as the key performance parameter can be monitored with a sensor or by periodic inspection. This approach greatly reduces the occurrence of sudden failures and unexpected shutdown or loss of availability of the system.

Figure 9.2 shows a degradation analysis plot. This illustrates measurements at intervals, declining values of a key parameter and the acceptable limit threshold that defines failure of the unit.

There are useful synergies between diagnostics/prognostics and HALT, HASS, HASS, AST and ALT. During the development stage, HALT or AST can be used to accelerate a prognostic experiment and development of diagnostic sensors and data capture. Once failure modes are discovered in HALT or AST, condition monitoring with sensors or built-in tests can be used to detect degradation of key performance parameters. The diagnostic capability can be evaluated in extended application of accelerated stress using HALT or AST during product development. This accelerates the degradation and so the response of the sensors can be evaluated. After degradation is determined, the units can be analyzed to confirm location and damage similar to failures noted in HALT earlier. The diagnostics can also

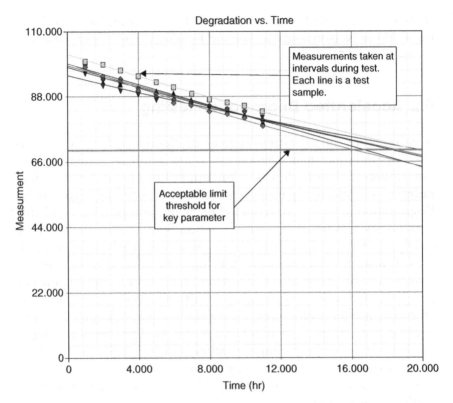

Figure 9.2 Degradation analysis plot with failure threshold [6]. Source: Data from http://www.reliawiki.org/index.php/Degradation_Data_Analysis

be useful for developing a proof of screen for HASS and provide confidence that HASS is not degrading life of the units shipped to the customer [3].

9.3 Accelerated Life Test Planning

Conducting a quantitative ALT begins with selecting the unit to be tested and the applicable stresses to be applied in the test. HALT or AST may have determined that particular assemblies in the product have less design margin or are more critical to the operation of the system. So these assemblies may need to complete ALT to demonstrate their ability to perform reliably at particular stress levels for a specified duration. Planning an ALT requires the following inputs:

1. What stresses are applied to the units?
   a. Potential issues and failure modes identified in FMEA or DRBFM
   b. Customer-provided application profiles – may require time-varying stress profiles based on the operational profile [5]
   c. Single stress or multiple stresses
2. What parameters will be measured during the test to monitor unit performance?
3. What is the anticipated stress level in the application including stress distributions and worst case time and duration?
4. What is the upper limit of the unit relative to the applied stress determined in HALT/AST?
5. Stress levels to be applied during the ALT?
   a. Single constant stress
   b. Two or three combined stresses such as temperature and humidity or temperature, humidity and vibration
   c. Time varying stress profile to emulate application but at accelerated stress levels [5]
6. How many test samples are needed?
7. How will the units be powered?
8. Data acquisition equipment needed to monitor unit performance and record status?
   a. Which operating parameters to be monitored to determine system status and health
   b. How key parameters will be measured
   c. Sampling rate and interval to record data
   d. Data storage capacity needed
   e. Stress levels or profile applied
9. Test equipment needed to apply the selected stresses to the product?
   a. Test chambers, programmable power supplies, vibration shakers, mechanical support fixtures and sensors required
   b. Loads and loading profiles required
   c. Environment conditions during test: environmental stresses needed and environmental conditions maintained independent of test stresses applied

Table 9.1 ALT test plan table to capture key factors

Test Run	Stress high	Stress med	Stress low	Samples tested	Number of cycles to failure	Time to failure	Failure mode
ALT 1	°C max, VAC ALT 1 profile						
ALT 2		°C max, VAC ALT 2 profile					
ALT 3			°C, max, VAC ALT 3 profile				

10. Required test schedule to support product development?
11. Location and resources assigned to the test or arrangements with outside laboratory qualified to perform accelerated testing?

The test plan information should be reviewed and agreed upon by the project team. Once consensus is reached, the resulting test plan should be documented in the format required for the organization. With the test plan in place, acquiring the required equipment and setting up the test can begin. Sufficient test samples are ordered, data acquisition and operating equipment to power and control the units under test and monitor selected parameters is assigned or acquired.

A table such as the one shown in Table 9.1 can be used to capture test planning information in a concise format. This can be included in a test plan document formatted per internal or industry standards to more completely describe the selection of stresses, equipment used, data collection, test samples and data analysis. Including test data and analysis results with the plan forms a single document to capture what was learned in the testing.

9.4 Pitfalls of Accelerated Life Testing

There are numerous potential pitfalls in planning and executing accelerated life tests. Determining the appropriate accelerating variables and the associated life–stress model adequate for extrapolation are critical concerns. Over-acceleration of applied stresses leads to failure modes that are not representative of the field application of the product. Other pitfalls include:

- equal test unit allocation at all levels of the accelerating stress
- using an incorrect stress acceleration relationship
- use of inaccurate activation energy values such as those for materials from references
- attempting quantitative accelerated life test at the system level with incorrect acceleration levels, insufficient samples at each stress level and not considering stress interactions
- insufficient number of failures at all stress levels
- testing at overly high stress levels causing new failure modes and applying too much acceleration to the analysis

- considering only obvious failure modes and not completing a complete failure analysis on all failed samples
- not using good degradation data to monitor units and to provide useful data points for analysis [4].

It is essential to consider these pitfalls during planning of accelerated life tests. Pressure from management to reduce test time and cost can cause decisions that attempt to disregard these pitfalls during test planning or running the test. Doing so will compromise the results of the test and produce misleading conclusions that lead to unexpected failures of the product in the field. Used carefully at the appropriate level of assembly with carefully selected stresses and stress levels, quantitative accelerated life test can be useful during product development and can provide confidence in design choices.

9.5 Analysis Considerations

As test failure data becomes available during quantitative accelerated life test, analysis of the data can begin using software tools or manual calculations to produce interim results. These can be used to check if the test is progressing as expected or if unforeseen problems are occurring. Are the failure modes as expected from risk analysis, AST and HALT? Is degradation in key performance parameters being measured?

Analysis should follow the test plan stress levels and application use level stress and use the life–stress relationship anticipated based on expected failure mechanisms. Times or cycles to failure, failure symptoms and measurements and observations must be recorded.

When failures have occurred at a minimum of two of the accelerated test levels, times to failure and run times of surviving samples can be entered into software applications such as Reliasoft ALTA, Minitab or spreadsheet tools to check life–stress relationships and initial indications of extrapolated life or reliability. As the test progresses and additional failures occur, the analysis can be updated. There must be failures at each accelerated stress level to complete the analysis. Not all samples need to fail, but test run times on remaining samples need to be recorded and used in the analysis.

Additional background and guidance on analysis of quantitative accelerated life test data can be found in the references listed below or in the Reliasoft e-text at: http://reliawiki.org/index.php/Accelerated_Life_Testing_Data_Analysis_Reference

Bibliography

[1] Nelson, W. (190, 2004) Accelerated Testing Statistical Models, Test Plans, and Data Analysis, John Wiley & Sons, Inc., Hoboken, New Jersey.
[2] Meeker, W. and Escobar, L. (1998) Statistical Methods for Reliability Data, John Wiley & Sons, New York.
[3] Silverman, M. and Hofmeister, J., (2012), The Useful Synergies Between Prognostics and HALT and HASS, Proceedings of The Reliability and Maintainability Symposium (RAMS), January 23–26, 2012, Reno, NV, SAE International, Warrendale, PA.
[4] Sarakakis, G., Meeker, W. and Gerokostopoulos, A. (2012) Pitfalls in Conducting and Interpreting the Results of Accelerated Tests, Proceedings of the Applied Reliability Symposium, June 13–15, 2012, New Orleans, LA, ARS, Reliasoft, Tucson, AZ.
[5] Paschkewitz, J. (2010), Calibrated Accelerated Test (CALT) with Time Varying Stress Profiles, The Reliability Edge, Volume 10, Issue 1, 16–21.
[6] Reliasoft, ALTA Example 3 – Accelerated Degradation, http://www.reliasoft.com/alta/examples/rc3/index.htm (accessed 21 September 2015).

10

Failure Analysis and Corrective Action

Testing to failure and understanding the physical mechanisms causing the failures is an essential part of product development and ensuring reliable products. Finding the design limits of components and assemblies is essential to make the product robust enough to withstand various stresses. However, the testing methods described previously only generate the failures. The essential next step is determining the physical cause of the failures and taking the needed corrective actions to prevent these failures from occurring in the customer application for the product.

10.1 Failure Analysis and Knowledge Capture

Analyzing test failures to understand the mechanisms that have caused them is an essential part of evaluating test results. Another important aspect is knowledge capture and reuse so the learning is available to follow-on project teams. Failure analysis and reporting accomplishes this objective and should be applied at all phases of development. An established failure analysis process tailored to the type of products made by the organization improves reliability, increases customer

Next Generation HALT and HASS: Robust Design of Electronics and Systems, First Edition.
Kirk A. Gray and John J. Paschkewitz.
© 2016 John Wiley & Sons, Ltd. Published 2016 by John Wiley & Sons, Ltd.

satisfaction and lowers overall product cost. It enables root cause determination and leads to faster resolution of product weaknesses and production issues. Failure analysis is a team effort. It requires active management support and the teamwork of experts in design, testing, supply management, manufacturing and quality. All of the team members contribute their expertise to isolating and correcting failure mechanisms.

Failure analysis is a progressive process that begins with documenting failure mode characteristics and observations, continues with non-destructive methods to examine failed parts, and finally dissection to expose failure damage and indicate failure mechanisms.

- Basic information collection
  - recovery of failed samples
  - electrical test, microscopy, digital photography
  - document initial evidence of the failure mode for further examination
- Non-destructive methods
  - X-ray reveals internal damage at failure site without disturbing physical evidence
- Disassembly/de-capsulation
  - tools or chemicals to remove layers and expose evidence of failure mechanism
- Scanning electron microscopy, energy dispersive spectroscopy
  - defects, corrosion, contamination, material failure identified
- Acoustic microscopy, imaging (voids and defects)

There are important decisions required in developing a failure analysis lab. Planning begins with an understanding of the types of failures typically experienced with the product. Acquiring capability for the failure analysis lab is based on the types of equipment needed to determine the failure mechanisms most frequently experienced. Most labs typically have basic imaging and measurement capabilities to characterize and capture failure evidence and this includes high resolution digital imaging, microscopy and electrical or dimensional measurement equipment. At the next level, non-destructive testing equipment such as X-ray systems and infrared imaging cameras are often used and are a more expensive acquisition. Beyond this, even more sophisticated equipment such as scanning electron micro-scopes, energy dispersive spectroscopy and acoustic microscopy are considerably more expensive and are justified only if use is frequent enough to have them in the lab. If not, this type of analysis can be sent to outside independent labs equipped to perform these investigations.

Knowledge capture and reuse enables follow-on project teams to easily access what was learned in previous projects and use it as a baseline for derivative projects or improvements to extend product life or application. The data items should be easily searchable and retained in an organization-wide tool to facilitate locating and using the information.

10.2 Review of Test Results and Failure Analysis

HALT, AST and ALT all are tests to failure. Failure analysis as described above helps reveal the physical failure mechanisms so they can be understood. After test results and failure analysis exami-nation are available, the third phase of GD^3 is conducted using Design Review Based on Test Results (DRBTR). This is similar to the DRBFM used to anticipate risks and failure modes at the start of the project, but now the team led by the test engineer documents the results and compares them to anticipated results and results of pre-vious tests on similar or predecessor products. Surviving and failed test samples are dissected and available for examination by the reviewers. Concerns and inputs on failure mechanisms are collected during the review and documented in the DRBTR form for knowledge capture. This format is shown broken into three sections in Figure 10.1. Figure 10.2 shows the Robustness Indicator Figure described earlier. It is useful for showing test results during a DRBTR.

As in the earlier DRBFM for risk and potential problem identification, the key part of the DRBTR is a focus on actions needed to resolve any remaining problems. The team conducting the DRBTR also establishes the priority of corrective actions based on the impact to the project and the customer, the schedule impacts and the cost to complete. The actions are tracked to closure by the project team as the product development moves forward (5,6).

Test results and test engineering inputs:

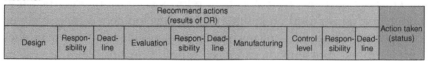

Item or part	Characteristic or parameter tested	Comparison of test result with expected result or previous results	Causes of test results/failure	
			Probable or confirmed cause of test results/ failure	Reviewer inputs related to cause

Effects and review comments:

Effect of failure on customer (possible progression of events to customer complaint)	Priority	Summary of test results review comments		
		Design	Test and evaluation	Manufacturing

Actions:

Recommend actions (results of DR)											Action taken (status)
Design	Respon-sibility	Dead-line	Evaluation	Respon-sibility	Dead-line	Manufacturing	Control level	Respon-sibility	Dead-line		

Figure 10.1 Design review by failure modes (DRBTR) results format [5]. Source: Haughey, 2012. Reproduced with permission of SAE

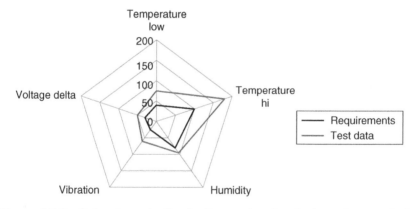

Figure 10.2 Robustness indicator figure showing test results margin for DRBTR discussion

10.3 Capture Test and Failure Analysis Results for Access on Follow-on Projects

Knowledge captured in previous projects is evaluated during the concept stage. Learning in design of experiments on components and materials can help isolate the causes of failure in testing or field returns.

The results of testing, data analysis and failure analysis reports and discussion during the DRBTR should be captured in a searchable tool for access by future project teams. This can be done using commercial software applications or using a tool like Microsoft Sharepoint to make documents such as test reports, failure analysis and the results of actions taken accessible on an organizational intranet site.

10.4 Analyzing Production and Field Return Failures

After the product is launched and production has started, failures detected in screening such as HASS and field returns should undergo the same failure analysis as used on failures during development testing. New or unexpected failure mechanisms highlight the need to examine manufacturing process problems or variation, supplier quality issues or system interface issues that were not detected during development testing.

The results of ongoing reliability tests such as HASS and re-HALT of units from the production line can be compared with field returns to determine if issues recur during production or if new issues appear that may be related to production process variation or changes in parts or materials by suppliers.

Bibliography

[1] Fleisher, J. (2015) Electronic Part Failure Analysis Tools and Techniques, Proceedings of The Reliability and Maintainability Symposium (RAMS), 26–29 January 2015, Palm Harbor, Florida.

[2] Martin, P.L. Electronic Failure Analysis Handbook, McGraw-Hill Companies, Inc.

[3] Paquette, T. Detective Work Finds Board Failure. Test & Measurement World (Aug 2006), pp.29–42.

[4] Tulkoff, C. and McLeish, J. (2014) Root Cause Analysis (RCA), Proceedings of The Reliability and Maintainability Symposium (RAMS), 27–30 January 2014, Colorado Springs, Colorado.

[5] Haughey, B. (2012) Design Review Based on Failure Modes (DRBFM) and Design Review Based on Test Results (DRBTR) Process Guidebook, SAE International, Warrendale, Pennsylvania USA.

[6] Adams, J.R. (1996) Failure Analysis System – Root Cause and Corrective Action, in Handbook of Reliability Engineering and Management, Second Edition, (eds. W.W. Ireson, C. Coombs, Jr., R. Moss), McGraw-Hill, New York, pp. 13.1–20.

11

Additional Applications of HALT Methods

11.1 Future of Reliability Engineering and HALT Methodology

This book was written to encourage a new perspective on reliability engineering and new applications. A change from probabilistic predictions to deterministic methods will focus reliability to address the most costly unreliability period for any company, the front portion of the classical bathtub curve, the declining hazard rate. Traditional probabilistic reliability engineering has done little to address early life failures, the latent defects that fail relatively shortly after use. The losses to a company for early life failures not only result in warranty replacement costs, but the much greater potential loss of market share from the perception of poor reliability.

In promoting the deterministic methods based on strength limits chasing irrelevant failure modes has been a common fear of those who have not performed HALT. Once a limit has been found, its relevance to reliability can be evaluated, but without finding limits there is no opportunity to improve the strength of the product, and the strength is the only value that can be known for sure when the frame of reference

Next Generation HALT and HASS: Robust Design of Electronics and Systems, First Edition.
Kirk A. Gray and John J. Paschkewitz.
© 2016 John Wiley & Sons, Ltd. Published 2016 by John Wiley & Sons, Ltd.

and metrics are based on the empirical strength limits of assemblies and systems. The methods called HALT and HASS were the first methods to apply this orientation and frame of reference.

To adopt the new orientation to empirical stress limits of systems and to develop new reliability discriminators from it, we must abandon the strong belief that reliability is a predictable and controllable variable to which accurate and beneficial models can be applied. A limited number of electronics devices do have intrinsic degradation mechanisms, such as battery technologies and Aluminum Electrolytic capacitors, that modeling and prediction provide the necessary useful life. For the vast majority of electric components, trying to quantify the intrinsic life entitlement and wear-out modes in order to predict reliability is irrelevant, as most will be retired before the intrinsic life is consumed. Quantifying an electronic system's life is similar to quantifying the 'life' of a newly constructed house. It will have much more life in most cases than will ever be needed.

The change in orientation from probabilistic and statistical predictions to empirical deterministic testing with HALT methods will continue to be controversial as long as reliability engineers do not discover how significant the strength of their designs can be with only standard materials and methods.

So many of the real causes of early life unreliability, such as low design margins and manufacturing flaws are overlooked, and so most efforts to improve reliability should be focused on early discovery and elimination of design and manufacturing errors. When reliability engineers begin to focus on finding the exceptions to good design and manufacturing, which result in products being unreliable, and capitalize on the full strength of materials and operational variables to be used as reliability metrics there will be new opportunities to build more sensitive and relevant discriminators to detect process excursions and to measure process variability comprehensively for a system.

A large portion of traditional reliability engineering activities for electronics are performed during the design phase, before the first hardware prototypes are available. In HALT and HASS, the majority of activities are focused on testing and analysis when hardware becomes available. HALT is not a singular test and should be applied for each hardware phase until market release. If it has been determined that HASS processes are justified for the product, then HASS should be begun. If the engineering changes (EC) are made to the original product they should be evaluated to determine if they could affect the stress margins. If a proposed EC change is judged to be a risk of reducing the stress/strength margins that were established in the HALT of the original design and used to define the stresses used for HASS, then HALT should be applied to verify that the same or better stress/strength margins are present after the EC has been implemented.

11.2 Winning the Hearts and Minds of the HALT Skeptics

Almost all companies new to HALT methods will have many skeptical engineers that will need to be convinced that HALT is a valuable and relevant method for finding latent design flaws, and that HASS is relevant for finding latent manufacturing flaws. The concept that relevant failure mechanisms can be observed at stresses that are well beyond end use environments is especially difficult for those from a traditional reliability engineering background which focused on quantitative reliability predictions. Getting from the point of 'HALT and HASS stresses broke good hardware' to 'HALT and HASS found a reliability improvement opportunity' for many engineers who are not familiar with the radical change in orientation and frame of reference can be a long journey. There are several paths that can help.

11.2.1 Analysis of Field Failures

An excellent way to win the hearts and minds of HALT skeptics and to demonstrate the benefits of HALT is to apply HALT and HASS to a product with known reliability weaknesses. Many of the design weaknesses and latent defects that were manifested as field failures can be detected in high stress conditions. If the failure mechanism from a design is known then showing that HALT or HASS would have found the reliability issues that resulted in warranty costs can be very convincing, and if the warranty costs are known then it can help to make a business case for HALT chambers and supporting equipment.

Some of the causes of field failures are easily demonstrated or realized to be able to be detected with HALT stresses. Loose cable connections, poorly seated components and adjacent components shorting after contact would all logically be stimulated from latent defect to a patent failure under random vibration. Cracks in traces, conductive vias or solder joints can be causes of intermittent failures that require more intrusive failure analysis to discover, but once they have been discovered it can be logically shown that there is a high probability of precipitation and detection if the system had been monitored under vibration and or thermal cycling.

11.3 Test of No Fault Found Units

From 20% to 70% of parts and system returned to manufacturers are found to operate normally when tested by the manufacturer (1). Since no failure is identified when they are tested they are classified as no failure found (NFF) or cannot duplicate (CND). In many of the returned units declared NFF, there may be a marginal condition such as a loose connector, or an active component that is so near to a margin limit that a slight change in temperature or voltage produces a failure. One factor that may lead to an NFF is that the bench testing of the returned unit may not include end use conditions that the product was operating in.

11.4 HALT for Reliable Supplier Selection

HALT is the most effective tool for reliability development of electronic systems, but it cannot be performed before there is hardware to test. A classic question from traditional reliability engineers is 'What reliability engineering work can be done during the design phase before the system is available to test?'

During the design stage, reliability engineers should be part of design reviews to help find and prevent overlooked issues in hardware layouts that would end up becoming system weaknesses. FMEA of systems designs can help prioritize the HALT on the different subsystems, levels of assembly and functions that will be most valuable to apply during HALT.

Reliability engineers should always be seeking to understand why real hardware fails, what are the real causes of unreliability and be reviewing product return failure analysis details. Only by observing the causes of unreliability can reliability in new products be increased.

When the industry begins to move away from spending resources on quantitative predictions (MTBF) for systems, it will allow many more hours and resources be applied to better planning and applications of HALT on subsystems and different hardware configurations, and to the development of higher resolution of discriminating measurements of operational reliability while under stress conditions.

Reliability engineers should also be working with the procurement department by reviewing and evaluating a potential subsystem supplier's reliability development processes. If a supplier claims to use HALT and HASS, those claims should be considered with caution. HALT and HASS methods are becoming more prevalent in the electronics industry but many may still claim to perform HALT more for marketing promotion than for actual product development. It has been the author's discovery that many who claim to perform HALT do the HALT but do not actually make changes to the designs based on what weaknesses are found in HALT. It is relatively easy to stress products to failures, and much harder to have design engineers agree to and implement changes to a product that has failed for stresses well beyond environmental design specifications. For those companies, the value of HALT is mostly in claiming to use the methods.

The quality of the supplier's HALT process cannot be known unless the supplier is willing to provide the HALT test procedure and detailed decisions on a system and what weaknesses, if found in HALT, were improved or for what reasons it was not. Most suppliers will not provide detailed HALT reports just as they would not release other new product development tests.

As the speed of electronic systems innovation increases electronics design and manufacturing companies are in need of faster reliability qualification of subsystems used in the total system before market release. Using the knowledge that in most cases the larger the empirical stress capability of the design of a subsystem, the higher the potential reliability, HALT can be used to compare the stress limits and capabilities of available subsystems to compare their potential reliability before incorporating them into a new design.

Of course there are two key ingredients necessary to produce a reliable electronic system: a robust design along with capable and consistent manufacturing processes. Any robust design can have poor reliability if it is poorly manufactured and there are many opportunities in the multiple levels of manufacturing systems for latent defects to be introduced.

During the design and development of a new electronic system by an electronics OEM there are subsystems, such as power supplies, that have more than one potential supplier to choose from. Once the technical specifications and cost requirements have been met for the new system design, HALT can help provide relatively rapid strength and stress margin information to compare multiple potential suppliers of a subsystem. 'Off the shelf' power supplies are generally available to evaluate and test early in the development cycle.

The OEM that will be incorporating it into their system should perform HALT on all the samples with the same sequence of stresses, the same sample sizes and the same HALT chamber if possible.

HALT can also be used to aid in the selection of reliable second sources of components for high volume systems such as IT hardware. All component suppliers must have stress margins or guard bands to allow for sufficient yields within the manufacturing variations during fabrication to meet performance specifications. Margins vary between suppliers, but how much margin in a component is up to the supplier and generally is not known to the user. UOL and LOL limits in HALT evaluations and their deviation may be a good discriminator of future reliability.

11.5 Comparisons of Stress Limits for Reliability Assessments

If the field of reliability engineering of electronics moves away from believing that reliability entitlements of electronics with no moving parts can be predicted, more resources can be used for using and applying deterministic limit approaches to reliability development.

If there are multiple suppliers that can provide the required subsystem at a competitive price, HALT may provide a relevant comparison of robustness and therefore reliability potential for the different supplier options. A simple relationship that might be considered common knowledge is that the greater the inherent strength of a system the more inherently reliable it will be. As shown in the Cisco analysis of thermal margins and warranty returns (Chapter 2), and as from a general understanding of stress and strength, the higher the strength of a subsystem the higher the reliability. Using HALT to compare strength limits using a variety of relevant stresses is a relatively quick method of providing a potential assessment of each supplier designed and manufactured strength.

Power conversion subsystems (power supplies) are an Achilles heel for most electronic systems, because without power all systems fail. HALT of power supplies is a very beneficial process in the fact that they are more likely to contain higher-mass components such as transformers and large value capacitors that can lead to mechanical bond and component adjacency spacing issues, as well as cooling issues.

Even though many power supply producers use HALT methods for development, the HALT process can be very different between suppliers. If using HALT for supplier selection of the highest strength system, the HALT procedure and conditions should be consistent for each supplier. For this reason, the HALT for all potential power supplies should be performed at the same HALT lab or in-house HALT lab under the same airflow, loading, vibration table and mechanical fixtures.

Although materials and methods of fabrication of electronics are changing and will change in the future, the introduction of latent defects causing field failures will be a challenge to creating a reliable electronic system. End use environments will always induce fatigue damage and aging of materials that will lead to latent defects becoming field failures in time. Because of this, stress testing to find latent defects will continue to be a very relevant and efficient tool for reliability development, especially if applied to the product's empirical operating limits.

Environmental stress drives the physical degradation of strength in materials and bonds, and eventually acts to consume the life entitlement of electronic systems. Therefore, stress testing to empirical limits for the process of discovery of potential reliability weaknesses will continue to be relevant for comparisons in future electronics materials and manufacturing technologies. Some electronic technologies with limited life, such as batteries, will need to be modeled and have future life projections known for reliable operation of systems using batteries. There is always the possibility that as electronics manufacturing changes in the future causing an intrinsic failure mechanism with a much shorter life than required for the applications will be created. As with any change that introduces a weakness, once the physics is understood it can be designed to extend the degradation rate so that the mechanism does not contribute to failures during the required life.

The continuing increase in IC density and lower voltages will make devices more sensitive to process variations that impact device parameters. Variation of device parameters inevitably results in variation of overall circuit parameters, affecting its performance and power consumption as well as decreasing the yield. Process, voltage and temperature (PVT) variation effects are especially pronounced in high speed systems, where requirements on performance parameters are stringent. PVT variation of key parameters in IC components has always been an important problem (2). Keeping up with Moore's law, by constantly shrinking the characteristic dimensions of IC components makes this problem more significant. In deep submicron processes variation of transistor threshold voltage can be over 30% and sheet resistance variation of poly-silicone resistors reaches 40% for some technologies (2).

11.6 Multiple Stress Limit Boundary Maps

Personal computers, server motherboards and digital microprocessor systems can have a significant HALT testing benefit when better built in control and monitoring of voltage and frequency for stress margining to find reliability risks. The PC motherboard uses many voltage regulators to supply various DC voltages throughout the system. It is difficult to manually add test access points for control and monitoring of the various outputs from the various voltage regulators in the circuit. Allowing for test interfaces to give access to hardware or software for controlling the voltage regulator outputs will make it much easier to quickly find empirical operational stress boundaries.

A typical design validation test performed on electronic systems during development is what is termed as a 'four corner test' of system

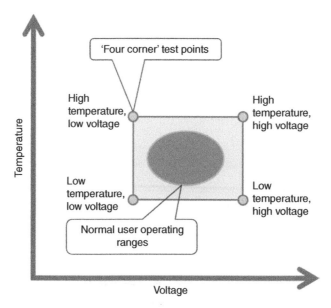

Figure 11.1 Temperature and voltage four corner test

voltages and temperature. The four corners refer to combinations of predetermined levels of high and low temperature combined with high and low voltages all within the design specifications of the circuit under test. It is used to verify the ability of the circuit to perform properly with variations of end use voltage and temperature. A graphic illustrating the test conditions is shown in Figure 11.1. The four corners the product is operated with are

1. high temperature, low voltage
2. high temperature, high voltage
3. low temperature, low voltage
4. low temperature, high voltage.

The four corner test described above is typically run on a small sample population and only provides pass/fail attribute data.

Applying the HALT philosophy to temperature and voltage, or any multiple stress test, and extending the stress to the point of operational limits would provide much more useful variable limit

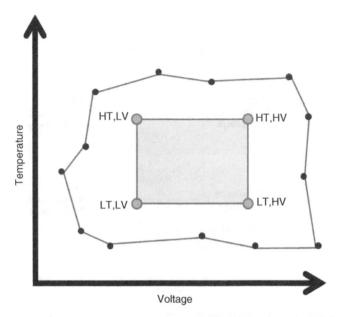

Figure 11.2 Voltages and temperature empirical operational boundaries

data. Using multiple stresses for graphical analysis and comparisons can be illustrated with an example of a two-dimensional graphical analysis of the interacting stresses of voltage and temperature in a digital system.

If the HALT philosophy and methodology were to be used for the temperature and voltage testing, the voltage and temperature increments would be smaller stepwise combinations and the stress applications would be extended to the point of the system failing to operate. Testing to empirical operational limits would result in an operational stress boundary map. A hypothetical operational stress boundary map for one sample is shown in Figure 11.2.

If the measurement is performed on many samples and compared, the distribution in the empirical limit boundaries which are the variations in operational capability or strength become observable in the graph. Considering that the end use voltage and temperature stress conditions are the load and that the empirical limit boundary is the strength, multiple measurements of multiple samples results in a

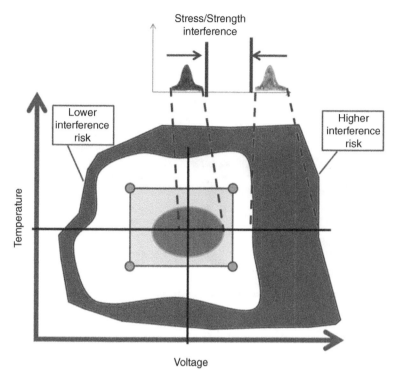

Figure 11.3 Empirical operational boundaries showing stress/strength distributions

two-dimensional stress/strength map showing variable margins. The widest distributions of strength indicate the highest risks for the stress to equal strength, and potential operational failures. Investigation into the cause of the observed larger variations in strength, which may be due to component or assembly variations, can lead to understanding how to reduce the variation and therefore reduce the risk of the stress and strength curves intersecting resulting in failures. Figure 11.3 illustrates graphically the ability to identify stress combinations of potential risk. The reliability risk is not so much from a potential faulty voltage regulator not providing the specified voltage – which is a possibility – but instead from the stack-up of component fabrication up through circuit board and system assembly PVT.

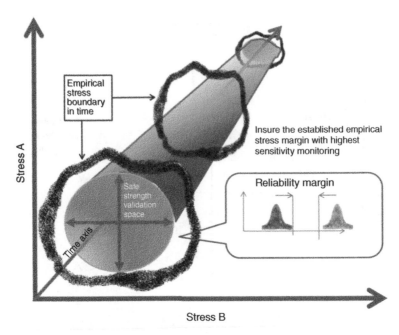

Figure 11.4 Two-dimensional safe stress margins for ongoing reliability monitoring

Empirical stress boundary limit graphic comparisons can be made for any combination of relevant stresses. In microprocessor based systems, varying the clock frequency, along with temperature and voltage would provide a three-dimensional boundary map which can be used for identifying PVT distributions that may cause operational reliability failures. The relevant stresses for operational boundary limits graphical comparisons is of course dependent on the product, the potential process variations and the end-use environment and applications.

Once the operational boundary maps are completed from the first hardware builds, they can be used for monitoring operational margins that, if reduced, may lead to poor reliability. The same stress boundary maps may also be useful for modeling performance degradation from aging mechanisms for system health prognostics. A generic stress graphic representation of the production monitoring or degradation monitoring is shown in Figure 11.4.

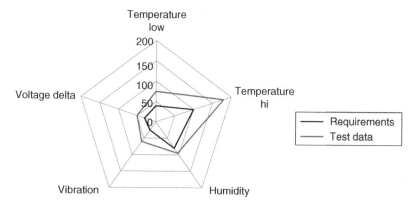

Figure 11.5 Robustness indicator figure

11.7 Robustness Indicator Figures

An alternative to stress boundary diagrams is to use a robustness indicator figure to show margin of the product relative to stresses applied. It shows both the required strength value and the strength value demonstrated in testing. It can be prepared using the radar chart feature in Microsoft Excel. Each stress is a spoke in the radar chart. The required, specified and test values are all shown on the same spoke. This visually shows the margin relative to each stress. A sample is shown in Figure 11.5.

11.8 Focusing on Deterministic Weakness Discovery Will Lead to New Tools

HALT is a simple concept, yet there is much more potential for the use of stress operational limits which sometimes are destruct limits, for reliability assurance and analysis. Advancing the use of HALT methods will occur with the deterministic measurement of the strength and performance of new electronics by applying more combinations of stress inputs while monitoring system performance. Because most electronics failures are due to errors in the multiple stages of design, manufacturing and assembly, building in the capability for variation and control of input stress variables such as voltage and clock frequency in digital systems will lead to higher resolution reliability discriminators for detecting latent manufacturing defects. The data derived from easier control and monitoring of voltage and frequency may lead to better graphical analysis using comparative boundary maps.

There is a continuing delusion in the field of reliability engineering that reliability in electronic systems with no moving parts can be predicted. The misdirection cannot be fought without shared evidence and knowledge of real field failures. Unfortunately the factors that prevent distribution and discussion of the causes of real field failures are not going to change anytime soon. Only when electronics reliability engineers seek to find and acknowledge the root cause of field failures in electronics products will they come to realize that rarely if ever are field failures the result of known intrinsic wear-out failure mechanisms. Additional supporting evidence of the power of HALT and HASS methods to find real reliability issues will come from those who effectively use HALT during product development and speak up for or publish the benefits.

As the benefits of HALT and HASS methods become more recognized and accepted as a key reliability task over reliability predictions, engineers will be able to focus on developing better ways to use stress and empirical stress limits and boundaries for better reliability assurance, assessment and discriminators.

11.9 Application of Limit Tests, AST and HALT Methodology to Products Other Than Electronics

A common misconception is that HALT can only be done in a HALT chamber and it only applies to electronics products. HALT is a methodology and is not equipment specific. The HALT/AST process can be applied to any product to determine empirical limits relative to a variety of applied stresses, to find product weaknesses and to improve design robustness.

The same process of anticipating loads and stresses applied to a product in the application, using a step stress accelerated test to find product limits relative to those stresses, and discovering design weaknesses applies to other electrical and mechanical products as well as electronics.

As described in Chapters 4, 5 and 6, the HALT process is essentially the same. The primary change for application to other products is the selection of stresses to be applied and applying those stresses to the product in the lab. These can be electrical loads with step increases of voltage or power, power fluctuation or transients applied to the product to find limits and weaknesses. Testing of electric heating elements is a good example of this type of application. Once the limits are found in a step stress test using increasing voltage, the limits are used to develop quantitative accelerated life tests as described in Chapter 9 to precipitate wear-out mechanisms such as oxidation or thermal fatigue of the heater elements.

Similarly, mechanical products can be limit tested by applying loads in steps to the design level and then up to overload conditions and failure, which can be yield or fracture. Similarly, resistance to wear can be tested in an abrasion tester and the unit can be inspected at intervals to determine its limit at degradation to an unacceptable level, which is the product or material's limit relative to abrasion loading. Hydraulic components can be subjected to step stress testing on a hydraulic test bench using increasing pressure. Hydraulic pumps and motors can be cycled under load until temperature reduces oil viscosity, and pressure or power output degrades to unacceptable levels. Another example is testing the effects of cavitation on hydraulic components or cycling pressures to produce seal leakage.

Corrosion is another example. A highly accelerated corrosion test (HACT) has been developed by Delta in Denmark. Based on the HALT methodology, it accelerates corrosion in a chamber developed to produce accelerated corrosive exposure and drying cycles. The results have been compared to samples exposed to field conditions. Similar levels of corrosion are correlated between field and test chamber exposure. See madebydelta.com for additional information on HACT.

These approaches illustrate how the HALT methodology and accelerated stress test can be applied to determine product design limits relative to various electrical, mechanical and corrosive stresses. This approach also identifies and confirms failure modes and product weaknesses for corrective action to produce more robust designs.

Bibliography

[1] No-fault-found and intermittent failures in Electronic Products. Qi, H., Ganesan, S. and Pecht, M. Microelectronics Reliability, Vol. 48, pp. 663–674.

[2] Process-Voltage-Temperature Variation Detection and Cancellation Using On-Chip Phase-Locked Loop. Melikyan, V.Sh., Durgaryan, A.A., Balabanyan, A.H., Balayan, E.H., Stanojlovic, M., Harutyunyan., A.G. Zlatibor: s.n., 2012. Proceeding 56th ETRAN Conference.

[3] Assessment of Reliability Concerns for Wide-Temperature Operation of Semiconductor Device Circuits. Kopanksi, J., Blackburn, D.L., Harman, G.G., Berning, D.W. Albuquerque, NM: Transactions of the First Internations High Temperature Electronics Conference, 1991.

Appendix: HALT and Reliability Case Histories

The world of reliability engineering and the increased understanding and improvement of the field of electronic and electromechanical reliability methods have a fundamental and significant blockage, which is the availability and dissemination of the underlying causes of warranty returns and reliability failures. It is rare to have any data on any reliability or methods of reliability development, as the liability and risk of disclosure in competitive industry will always exist and therefore will always be a serious limiting factor. Nothing can change in the discipline of reliability engineering of new products without real data on real failures in real products being disclosed and discussed within the profession.

Many engineers who have been involved in failures analysis of returned electronic systems are aware that most failures in the early years of electronics are due to mistakes in either the design, manufacturing or applications and finding errors through HALT methods should be the highest priority of any electronics design and manufacturing company.

Next Generation HALT and HASS: Robust Design of Electronics and Systems, First Edition.
Kirk A. Gray and John J. Paschkewitz.
© 2016 John Wiley & Sons, Ltd. Published 2016 by John Wiley & Sons, Ltd.

A.1 HALT Program at Space Systems Loral

Brian Kosinski and Dennis Cronin
Space Systems/Loral, Palo Alto, California, USA

A.1.1 Introduction

The suppliers of most US commercial satellites qualify hardware to tailored versions of Mil-STD-1540E and, for European suppliers, ECSS-E-10-03A, which has similar qualification test margins and environmental requirements. The philosophy behind both of these documents is to test units to predefined margins over flight predictions, for launch vibration and on-orbit thermal excursions, for example. This approach has served Space Systems/Loral (SS/L) well, as evidenced by excellent performance in the first year of operation, the time period monitored for robustness against infant mortality issues. Occasional qualification and acceptance test program escapes, however, adversely affect cost and schedule prior to launch. In this highly competitive market it is important to drive the number of failures after unit-level qualification as low as possible, with the ultimate goal of zero. The occasional qualification and acceptance test programs escapes are at least partially due to the fact that testing to flight predictions with some amount of margin does not fully protect against the statistical nature of design and manufacturing tolerances or of combinations of statistical variations. Figure A1.1 shows that high and low 'tails' of normal stress and strength distribution curves can overlap to produce an area of unreliability where failures can occur for any flight hardware unit type over its production life cycle.

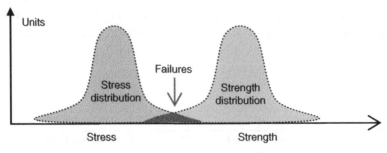

Figure A1.1 Statistical nature of stress vs. strength

A.1.2 Unreliability

Another way to state this point is that even if a unit passes 1540E or E-10-03A qualification, the margin remaining to failure is not known. The unit may be close to a limit that might not manifest itself in a failure until several production units are built and tested. Probabilities of not finding issues during qualification and acceptance temperature testing are illustrated in Figure A1.2.

Finding issues on the ground prior to launch can be painful from the satellite manufacturer's cost and schedule standpoint. However, finding an issue on orbit is far worse. It can severely damage a customer's business plan, result in the loss of a manufacturer's on-orbit incentive revenue and damage a manufacturer's reputation throughout the industry.

Since new technologies are frequently introduced to meet growing commercial satellite customer demands for power, bandwidth, pointing, etc., effective qualification test programs are essential. To reduce the number of design issues found in production after completion of MIL-STD-1540E qualifications and to reduce the probability of on-orbit failures, SS/L started performing highly accelerated life testing (HALT) on new development units in 1999.

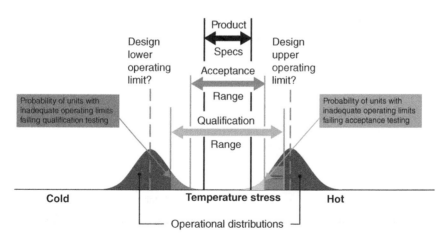

Figure A1.2 Probabilities of not finding issues during qualification and acceptance temperature testing

The question that many people ask when introduced to the HALT concept is whether HALT precipitates failures that will not occur in fielded units that operate in more benign and expected environments compared to the HALT stimulus. While this may sometimes be the case, the authors believe that in most cases HALT does identify real issues that can occur statistically over time. Numerous HALT papers published on commercial electronics indicate that failures induced in this manner accurately predict the failure modes that the product will encounter over time. One possible explanation for this – beyond pure stress versus strength statistics – is that strength decreases with age. Note in Figure A1.4 that the shifting of the strength statistical distribution to the left over time results in a similar amount of increased unreliability available to detect (the overlapping area of the two curves) as was previously shown in Figure A1.3, i.e. increasing test stresses and decreasing strength over time can produce similar results).

Thus the title 'highly accelerated life test' indicates that time to failure is 'accelerated' by increasing the test stress levels. As an example of how Figures A1.3 and A1.4 are applicable to the commercial satellite business, consider that it is typical for low-level assemblies to be vibration tested, then tested again at one or more higher levels of assembly. Each successive test adds some amount of cumulative fatigue stress, and then the unit must also survive the stresses induced by the launch prior to it performing its primary mission on orbit. The cumulative amount of vibration stress demonstrated in SS/L's HALT program is beyond what any unit will be subjected to during unit

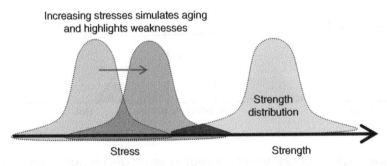

Figure A1.3 Stress testing principle. Source: Seusy, 1988

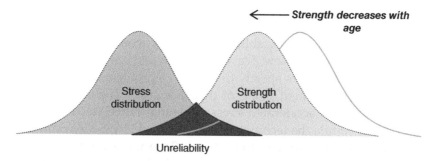

Strength decreases with age

Stress distribution

Strength distribution

Unreliability

Figure A1.4   Effect of time on strength. Source: Seusy, 1988

acceptance test and any subsequent stresses induced at higher levels of integration testing and launch.

By determining the root cause for each failure mode stimulated by HALT and implementing design changes to prevent their recurrence, product robustness, and thereby reliability, is improved. In other words, a successful HALT program will quickly create realistic equipment failures from which the designer learns root cause, potentially implements corrective action, and optimizes the design to push product limits out as far as possible. The fundamental differences in philosophy between MIL-spec qualification testing and HALT can be described as follows:

> *It may be typical for companies to hope that a unit passes the MIL-spec qualification test, or to explain away a failure, if one does occur, as an anomaly not to worry about. With HALT, however, the goal is to try to force failures, to understand product margins and identify weak links in the design in order to fix them and make the product more robust prior to moving into production*

A.2 Software Fault Isolation Using HALT and HASS[1]

The following paper from Allied Telesis (formerly Allied Telesyn) Labs in New Zealand illustrates how HALT and HASS can help detect software issues in digital systems.

---

[1] Written by Donovan Johnson, Senior Hardware and Reliability Test Engineer and Ken Franks, Hardware and Reliability Test Manager at Allied Telesis.

A.2.1 Introduction

In 2005, Allied Telesyn established a highly accelerated life testing (HALT) facility at their New Zealand research and development centre. This facility enables the speedy diagnosis of problems with products during the development phase of their life cycle. The company also undertakes highly accelerated stress screening (HASS) testing at their factory sites. While the techniques of HALT and HASS are commonly reported upon, it is difficult to find specific examples of faults found during testing. Allied Telesyn categorizes faults found during HALT and HASS by their root cause – either mechanical, electrical or software. This white paper focuses on the discovery of software failures in products and the isolation techniques involved.

HALT testing is conducted throughout the design stage of product development to highlight any major problems with products so they can be isolated, analyzed and corrected in a timely manner. HASS is conducted on a sample basis at the company's three factories. The screens created during HASS development are used for no trouble found (NTF) debugging, component qualification and software patch verification.

Until 2005, all of Allied Telesyn's HALT and HASS testing had been conducted offsite using an independent lab in San Jose, California. This testing process has been extremely useful for Allied Telesyn staff to gain experience and knowledge, but it has also been expensive because of the company's remote location in the South Island of New Zealand and the cost of travel to California.

In 2004, Dr Gregg Hobbs was invited to Allied Telesyn's New Zealand design centre to teach his 'Mastering HALT and HASS' seminar. This led to an experimental HALT conducted locally under the guidance of Dr Hobbs and ultimately to the implementation of a comprehensive HALT and HASS program in Christchurch. Allied Telesyn encountered multiple obstacles and challenges that are often inherent in setting up a HALT program. Initially, many engineers were skeptical about the relevance of failures found at the high stress levels applied during the HALT process. This response is common when first adjusting to overstress techniques, especially when testing is conducted remotely because it is difficult to directly isolate and diagnose software faults, let alone to debug from the other side of the world.

The cooperation of Allied Telesyn's Engineering team coupled with extensive feedback through comprehensive debug reports has overcome these initial frustrations. Top-level management support is an imperative for the implementation of any HALT and HASS program.

A.2.1.1 What is a Software Fault?

The term 'software fault' is defined at Allied Telesyn as a fault found in a product's:

- firmware, such as code in a programmable logic device (PLD)
- boot code, such as EPROM boot code
- operating system.

These software faults may occur because of changes in the performance of the associated hardware.

A.2.2 Testing and Monitoring

The sequence of tests applied to a product during HALT and HASS plays a critical role in the ability to uncover software faults. The monitoring process provides a snapshot of each product's status shortly before a failure occurs. The test sequence enables us to identify a myriad of software faults related to the fundamental operation of products, including clock signals, voltage rail monitoring and environmental factors, which can have an effect on product stability. Other factors that may also be monitored during testing include read/write timing, chip selects, reset pulses and signal integrity.

A considerable amount of debug information must be extracted from the testing process in order to highlight failures, without which fault isolation would be an extremely laborious process. The debug information may include, memory dumps, voltage measurement, clock observation, and other product specific measurements.

A.2.2.1 Testing and Monitoring for HALT

HALT testing and monitoring needs to be as comprehensive as possible. Experience shows that greater test coverage coupled with exhaustive product monitoring leads to a plethora of diagnostic information, which can then be used while debugging each failure mode. Test developers need to put considerable thought into developing a broad range of tests that will reveal the information required coupled with an all-encompassing monitoring program prior to starting the HALT.

A.2.2.2 Testing and Monitoring for HASS

Highly accelerated stress screening (HASS) occurs on multiple sample units of each product during mass manufacturing. Appropriate design measures are important for HASS, to ensure that external monitoring equipment can be utilized and should include the provisions for both hardware monitoring and built-in software testing in conjunction with data logging.

A.2.2.3 Typical Test and Monitoring Process

Allied Telesyn runs a 15 minute dwell at each step during the HALT process. During this time the unit is functionality tested and monitored for failure. Some examples of the testing conducted include:

1. external traffic test
   a. using industry standard equipment
2. power cycling
   a. voltage and frequency margining
3. internal memory test
   a. RAM test
   b. NVS test
4. internal packet generator test
   a. CPU test
   b. encryption engine test
   c. RAM test
5. other product-specific tests.

In addition to this test sequence the product is monitored in real time to ensure that the slightest change in operating specification is observed and recorded. The product monitoring includes:

1. voltage rails
2. frequency

3. temperature
4. critical system signals
5. self-diagnostics.

All of these exert some influence over whether a fault will be discovered and corrected or remain unnoticed and hamper a product for the duration of its life in the field.

A.2.2.4 Examples of Fault Isolation Techniques

HALT is a creative process with many paths to a reliable product and many innovative approaches to fault isolation. Using hardware and software tools, fault isolation can be as rudimentary or as complex as the engineer's creativity allows.

The following examples demonstrate how locating the specific cause of a failure can be identified in a simple and efficient manner.

(Note that during the application of thermal isolation it is important to monitor the temperature of the component or circuit using a thermocouple or similar device.)

A.2.3 Freeze Spray Example

The controlled application of freeze spray to a suspect component or circuit identifies the area where a failure is generated as illustrated if Figure A2.1. This keeps the temperature of the suspected cause of failure at room temperature while the rest of the product is heated, providing a straightforward and cost-effective way to isolate heat related faults.

A.2.4 Heat Application Example

The controlled application of heat to a suspect component via a power resistor or Peltier device identifies components that are failing due to cold temperature as shown in Figure A2.2. Keeping this component isolated from the surrounding environmental conditions can provide a simple way of identifying the root cause of a failure.

Note that, in both of these examples, only the location of the failure has been identified. The cause of failure must be determined before deciding upon an appropriate course of action. The combination of

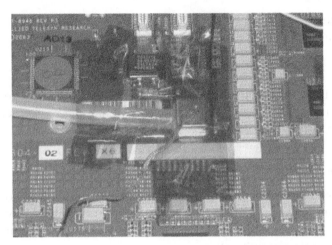

Figure A2.1   Freeze spray is applied to a suspect component

Figure A2.2   Power resistors are applied to a suspect circuit

testing, monitoring, fault isolation and effective debug information will, in most cases, lead to the root cause of the failure. Once the root cause is determined, the fault is analyzed and an appropriate corrective action implemented allowing the testing to continue.

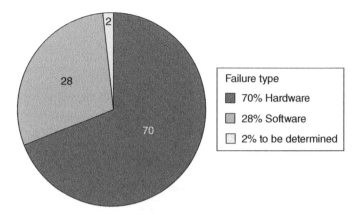

Figure A2.3 Fault type summary by percentage

A.2.5 Allied Telesyn Fault Examples

This section focuses on software failures found by Allied Telesyn during HALT and HASS testing. Figure A2.3 shows that almost one-third of all failures found during Allied Telesyn's HALT testing are attributed to software. Two percent of failures are yet to be determined.

Figure A2.4 compares the failure type to the stress applied, illustrating that all software related failures occur within the first three steps of the HALT process. In most cases, the problem is identified and eliminated prior to the product undergoing combined environment testing.

A.2.6 Common Software Faults Identified During HALT and HASS

The following examples demonstrate the variety of software failures discovered by Allied Telesyn during HALT and HASS.

A.2.6.1 Abnormal LED Activity

Light emitting diodes (LEDs) are used on network products such as routers and switches to indicate their current status. LEDs are often driven directly from the physical interface for basic functionality or, in more complex designs, driven through one or more programmable

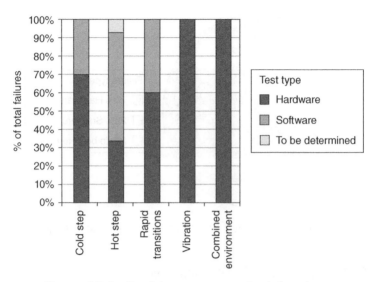

Figure A2.4 Fault type summary by failure type

logic devices (PLD). This logic includes specific timing signals relative to status display, reset signals and chip selection.

One fault occurred after a hard power cycle but would not occur after a soft reset. The failure caused random LED activity when powering up, which remained until either the temperature was raised or the soft reset activated. This anomaly was found during cold step testing at −10°C and attributed to the reset pulse timing inside the PLD code.

The fault was isolated by applying heat to the suspect component, and it led to a fix that was applied within 30 minutes. The mechanics of this failure lay in the code of the PLD. Under a hard power cycle the PLD reset pulse duration was insufficient, which led to the board powering up in an unknown state. After the PLD code was updated there were no further LED indication errors to temperatures as low as −50°C.

A similar Allied Telesyn product that had been in mass production for three years had also exhibited this fault during standard production testing. The fault had slipped through the traditional design qualification testing when Allied Telesyn did not have a HALT program in place.

A.2.6.2 Switch Tuning

Network switches are built around silicon switching components, which divert packets via hardware switching. Some components require tuning over an extended temperature range at various input voltages; one such device is the Marvell Prestera 98EX115D.

While developing a new switch product, it was necessary to tune the silicon switch to ensure correct operation over a broad range of temperatures. At the start of the HALT test, the unit had an upper operating limit (UOL) of 70°C and a lower operating limit (LOL) of −20°C.

Step stressing revealed that the switch was not tuned correctly, and over a few days new software versions were created and tested, ultimately fixing the errors. Corrective action through software enhancements culminated in a UOL of greater than 100°C with a LOL of less than −60°C.

A.2.6.3 Power-Up Sequencing

Power-up sequencing is a common requirement in today's complex electronics, and while there are many dedicated power sequencing components available, it is occasionally more efficient to use onboard logic to control the voltage rail sequencing.

A product that underwent HALT testing exhibited a failure where various components were not functional after a power cycle at −20°C. Applying thermal isolation techniques identified the PLD device as the faulty component. The code that controlled the power-up sequencing proved unreliable at low temperatures.

This problem had previously been identified and fixed during prototype development at Allied Telesyn's design centre. However, HALT revealed that the applied fix had simply shifted the failure, making it less replicable and this allowed Allied Telesyn to apply a more robust corrective action. The code inside the PLD was modified, allowing the product to reach temperatures lower than −50°C without failure.

A.2.6.4 System Crash

System crashes during the design phase are inevitable when developing prototypes that contain onboard CPUs. The key is to stimulate these crashes during development and allow a fix to be implemented at minimal resource and time cost.

A product that had been in the field for six months was taken to a HALT lab to investigate its operating margins. At the beginning of the HALT the product attained an UOL of 70°C, and a system crash was observed at this temperature. By changing a register setting for the initialization of a particular memory interface inside the boot code, the unit was able to function at temperatures above 100°C.

In addition to the software fault, a flaw within the CPU silicon was revealed, which amplified the effects of the software fault. The complete solution came in the release of a new revision of the CPU, coupled with the original boot code change.

The same problem eventually appeared on three separate products, two of which occurred in the field environment. No further failures occurred after the boot code was modified and the CPU flaw corrected.

A.2.6.5 System Silent Reboot

A silent reboot occurs when the product reboots without displaying any error or debug messages. These silent reboots are notoriously frustrating to debug and often require exhaustive troubleshooting.

A major Allied Telesyn customer with 22,000 units of one type of product in a major network was experiencing eight silent reboots each day, which represents a 0.036% failure rate. Allied Telesyn faced the problem of how to replicate a failure that only occurred on 0.036% of units. The same fault took weeks to replicate intermittently using traditional methods, which led to the use of a simplified HALT in an attempt to identify the problem. The same failure mode was repeatedly replicated in less than one day of testing, enabling software engineers to isolate and remedy the failure cause in a short space of time.

Rapid thermal transitions exposed a flaw in software during temperature ramps, even though the initial failure occurred in a moderate climate inside a server room. The failure mode was only apparent when running one particular test. A software patch was released to fix this problem.

A.2.7 Summary

Software is an integral part of today's electronics, and ensuring that the software on our products is reliable is a vital part of delivering quality products to our customers. Reliability failures in hardware are usually caused by wear. However, software does not wear out and may continue to function after an initial failure, making the fault harder to replicate, isolate and analyze. HALT and HASS are excellent tools for uncovering dormant defects in both hardware and software, but without an exhaustive test and monitoring plan, many software faults will continue to go undetected.

It is essential for a comprehensive HALT program to include relevant monitoring and test suites that uncover more than just hardware faults. Having a comprehensive test plan leads to substantial information that can then be used to judge the relevance of each failure and the corrective action required.

Experience at Allied Telesyn has shown that accurate fault isolation is a fundamental aspect of HALT and HASS testing, without which the process of analyzing faults and implementing corrective actions may be an exercise of trial and error. Thorough planning and consultation prior to conducting HALT should provide an overall picture of the product's dependencies and highlight the steps necessary for testing and monitoring these critical aspects. The HALT chamber is a very useful tool for increasing the reliability of a product, but it is not a magic box and will only provide the outlined results if used to its full capabilities.

A.3 Watlow HALT and HASS Application

Mark Wagner, Watlow Controls, Winona, MN

Watlow designs and manufactures industrial heaters, sensors and controllers – all of the components of a thermal system. Designing and manufacturing the complete thermal system allows Watlow to recommend, develop and deliver optimum thermal solutions for our customers' equipment and process heat requirements. Watlow has more than 93 years providing the most innovative thermal solutions to customers in a wide range of industries.

Watlow was founded in 1922, and is family owned. It has grown in product capability, market experience and global reach. Watlow holds more than 200 patents and employs over 2000 employees working in 12 manufacturing facilities in the United States, Mexico, Europe and Asia. It has sales offices in 14 countries around the world and a global distributor network.

Watlow began HALT testing in 2001 and has evolved the process since then. The modulated excitation profile has been refined, and voltage and power variation have been added as additional stresses applied to products. Software and firmware issues have been detected and corrected using HALT. The HALT process has been defined in an internal lab procedure for control electronics. The HALT methodology has been extended to heaters and sensors to find design limits used in quantitative accelerated life testing.

Example 1: HALT was used on a control PWB to determine limits and find weaknesses. The following results were noted: The lower operational temperature limit = −80°C (this was the lower temp limit of the chamber, not that a failure was seen). The vibration limit = 40Gs (this was the upper vibration limit of the chamber, not that a failure was seen). The upper operational temperature limit = 135°C. After the internal lab procedure effort using temperature dwells, temperature cycling and random vibration concluded, an additional stress of varying the supply voltage was attempted. It was found that increasing the supply above 18 V would result in a transistor failure. It was then determined to change that transistor to a more robust device.

Example 2: HALT was used to analyze field failure returns. Information was obtained from the returned unit description: 'When the device is powered up and temperature of the unit is brought down to below 0°C, if the unit is switched OFF at this temperature for at least one minute and powered up again, the firmware will hang.' The failure was replicated in the HALT chamber. Failure analysis indicated returned units passed electrical testing at room (25°C), hot (85°C) and cold (−40°C) temperatures using ATE. During failure analysis, errata from the microcontroller manufacturer were reviewed. None aligned with the failure mode, but one had a similar failure description and described a special writing command string to the flash memory. The errata write sequence was attempted and fixed failing units. This prompted discussions with the microcontroller manufacturer for long-term and short-term corrective actions. At some point in the microcontroller manufacturing, a die change was implemented which adversely affected the micro flash memory.

Example 3: HALT on a new product with a specified temperature range of −40°C to +85°C showed that the product operated correctly to −100°C, but began failing at 40°C and became inoperable at 50°C.

Discussions with firmware developers revealed there were multiple sources for code development. Internal communications architecture used a serial peripheral interface (SPI). Two SPI components inadvertently had their latch set to a rising edge condition. All other SPI components had their latch set to falling edge condition. Modified software corrected the problem.

Example 4: Watlow uses HASS and HASA to screen selected products in production. A stress profile shown in Figure A3.1 is used to precipitate and detect latent defects. The process uses a screen based on HALT results and a proof of screen (POS) process and a profile generated from the upper operating temperature limit (UOTL), lower operating temperature limit (LOTL) and operating vibration limit (OVL). The starting point is typically 15–20% less than UOTL, LOTL and OVL values determined in HALT. Five to ten units undergo the proposed screen 30 times. The units are functionally tested before and after POS. If the POS units pass the repeated screen profiles, the assumption is that a unit subjected to the screen one time has minimal life extracted from it and can be shipped to the customer.

A recommendation from Watlow's HALT and HASS experience is that monitoring of the UUT while subjected to stress is critical. Catching faults during stress is paramount to finding weak links of the design. Develop as much monitoring as costs, resources and schedules will allow. Application of HALT and HASS has improved reliability of electronic

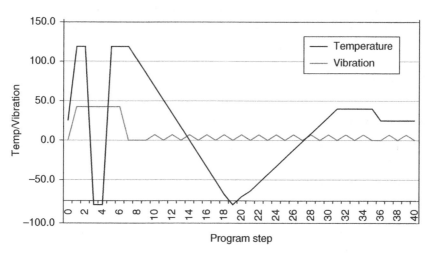

Figure A3.1 Typical HASS and HASA profile used at Watlow

controls. Numerous Weibull analyses have been completed on field data, and typical reliability values range from 97% to 99.5% at 3 years. HALT on new products and to verify field failures has been valuable in improving product reliability. HASS and HASA on selected products has improved production processes and reduced variability.

The HALT methodology has been extended to heater and sensor products by using step stress testing on heaters to find product limits and correct weaknesses prior to quantitative accelerated life tests. On heaters this is done by increasing power applied in steps and monitoring voltage, current and resistance until failure occurs as an open circuit or change in resistance. On sensors it can be done in a temperature chamber with increasing temperature cycle ranges until failure. Both of these accelerated stress test methods find limits of the product, determine failure modes and provide stress ranges for follow-on quantitative accelerated tests to extrapolate estimated life in the application at various expected usage case stress levels.

A.4 HALT and HASS Application in Electric Motor Control Electronics

After developing a HALT capability for several years, a system manufacturing OEM decided to evaluate the effectiveness of HALT and HASS in detecting and preventing relevant field failures. An example PWBA was tested in HALT as shown in the following figures.

Table A4.1 shows the order in which 10 board samples were subjected to HALT. The purpose of doing this is to vary the order of application of the stresses applied so that the order of application does not bias the results.

A.4.1 Example 1: HALT of a PWB assembly

The HALT test units were in cold temperature step stress, hot temperature step stress, cold vibration, hot vibration, temperature cycling and vibration and fast temperature cycling. The following figures show the profiles used in each test. Following that is a summary of the failures in each test phase and the failure modes identified.

The temperatures used in the cold step stress test are shown in Table A4.2. The air temperature applied to the sample, the chamber set point and measurement of the temperature on the sample are nearly

Table A4.1 Test sequence of ten sample modules

Sample #	1	2	3	4	5	6	7	8	9	10
HALT Test 1	Step Cold	RT Vibe	RT Vibe	Step Hot	Cold Vibe	RT Vibe	Hot Vibe	Fast Temp Cycle	Hot Vibe	Step Cold
HALT Test 2	Step Hot	Step Cold	Temp Cycle &Vibe	Repeat Step Hot	Hot Vibe	Temp Cycle & Vibe	Cold Vibe	Temp Cycle &Vibe	Cold Vibe	Temp Cycle &Vibe
HALT Test 3		Step Hot			Fast Temp Cycle				Fast Temp Cycle	

Table A4.2 Example of cold step stress test

	Temperature (°C)		
TIME (min)	Set point	Air	Sample
0	25	25	25
10	24	24	24
20	−45	−55	−40
30	−45	−45	−40
40	−55	−55	−50
50	−55	−55	−50
60	−65	−65	−60
70	−65	−65	−60
80	−75	−73	−70
90	−75	−74	−70
100	−80	−77	−75
110	−85	−83	−80
120	−96	−95	−85
130	−96	−95	−85
140	−103	−100	−95
155	−90	−95	−87

identical because of the direct way samples are heated and cooled in HALT. This helps achieve the fast ramp rates that are used in HALT. The results of the cold step stress portion of HALT are shown in Table A4.3. The operating and destruct limits were tested. There were

Table A4.3 Summary of cold step stress HALT results

Limit	Sample	Temperature (°C)	Status
	1	<=−90	No failure found
LOL	2	−90	Soft failures no communication
	10	−90	Soft failures no communication
	1	<=−100	No hard failures
LDL	2	<−100	No hard failures
	10	<−100	No hard failures

Cold step stress results

Table A4.4 Temperature Profile in Hot Step Stress HALT

Time (min)	Temperature (°C)		
	Set point	Air	Sample
0	25	25	25
10	25	25	25
20	60	58	55
40	75	76	75
50	75	75	75
60	80	80	80
80	90	90	90
90	90	90	90
100	100	100	100
110	100	100	100
120	115	115	115
130	115	115	115
140	120	121	121
150	120	120	120
160	132	130	130
170	130	130	130
180	146	145	144
190	125	145	145
200	153	152	150
210	150	150	150

soft failures at −90 to −100°C resulting in communication failures. These units recovered after the temperature was returned toward ambient. The test proceeded to the maximum cold temperature of the chamber but did not produce a hard or destruct failure. That is often the case in HALT, as extreme cold temperatures do not often produce destruct failures.

These step stress HALT tests determined lower operating and destruct limits.

Next, the high temperature HALT step stress test was completed. The temperature profile for this test is shown in Table A4.4. The results are shown in Table A4.5. There were recoverable failures at the Upper Operating Limit, but hard failures of relay functions at the Upper Destruct Limit.

These step stress HALT tests determined the upper operating and destruct limits and identified weak components and failure modes as indicated in Table A4.5.

Table A4.5   Results of hot step stress test of PWB modules

Hot step stress results			
Limit	Sample	Temperature (°C)	Status
UOL	1	130	Soft failures FETs + 1 fuse
	2	120	Soft failures FETs + 1 fuse
	4	110	Soft failures FETs
	4 retest	110	Soft failures FETs + 1 fuse
UDL	1	150	Soft failure, relay function, recovered near 130°C
	2	160	Communication failures recovered at lower temp
	4	150	Communication failure recovered at lower temp after cycle power on/off, 1 fuse failure
	4 retest	170	2 relay or related failures did not recover, multiple FET failures, 1 fuse failure, communication failure that did recover at lower temperature.

Next, the vibration step stress profile was applied to the PWBs. This involves a stepped increase in the Grms level of vibration applied to the samples.

After completion of the stepped stress testing, the test units were subjected to a combined temperature and vibration profile test profile as shown in Table A4.6. This combined stress profile often reveals weaknesses in the product design.

Failure modes found in the HALT of PWB modules were the following:

- mechanical failures: solder joint failures at some connector pins.
- electrical: probable relay failure and flickering of loads under hot vibration conditions.

Table A4.6 Vibration profile for combined temperature cycle and vibration HALT

	Combined temperature and vibration HALT			
	Temperature (°C)			Vibration Grms
Time (min)	Set point	Air	Sample	Set point
0	40	40	40	0
3	40	30	30	60
10	−50	0	0	60
15	60	−40	−40	60
20	60	70	50	60
27	−70	60	60	70
40	−70	−70	−55	70
42	90	−75	−70	70
50	90	100	75	70
55	−95	85	90	80
68	115	−100	−95	80
80	115	125	105	80
82	−95	100	100	80
92	−98	−98	−87	80
98	25	90	90	80
100	25	100	100	80
105	25	50	15	5
110	25	25	25	5
112	25	25	25	0
120	25	25	25	0

Table A4.7 Comparison of HALT and HASS results with field failure data

Component	HALT failures	HASS failures	Field failures
		Number of occurrences	
Capacitor 34, 39		1	1
Capacitor 4, 26		1	1
Rectifier		1	1
Contactor	1	1	1
Resistor		2	2
Diode		3	3
Capacitor 18		3	3
Q4	1	5	6
LEM	1		7
F1	1	1	7
IGBT	1	1	12

Recommendations include: failure modes need to be corrected if not cost prohibitive.

The outcome of the HALT testing was to apply the HALT methodology in combination with physics of failure reliability prediction methodologies and manufacturing process controls to continually improve the reliability of electric motor power and control electronic printed wiring boards and components.

A further study compared field failure results with HALT and HASS data on the same components. The results of that study are shown in Table A4.7. This shows that most of the field failures were revealed in HALT and HASS testing. The number of occurrences and the location of the failures by component for field, HALT and HASS are shown.

This study supports the use of HALT and HASS to reduce field failures on electronics printed wiring boards.

A.5 A HALT to HASS Case Study – Power Conversion Systems

A.5.1 An Efficient Path to a Company Adoption of HALT and HASS

The best way to illustrate the use of HALT and HASS and the warranty reduction benefit that can be realized from adapting the new paradigm is to present a case study from a company that has significant results.

A manufacturer in the mid 1990s was looking for more efficient and effective reliability development of their power conversion products and had heard of HALT and HASS processes.

A.5.2 Management Education

The path that the manufacturer followed to adoption of HALT and HASS methods was why the program became so successful for them. It began with a discussion of the methods with the reliability department manager and then an hour long one-on-one dialog with the CEO. They had a good understanding of the causes of unreliability in their products. They had detailed documentation of the failure mechanisms, which at the time was mainly various workmanship errors. Management recognized that many of the failures could have likely been detected during vibration and thermal cycling, and therefore there was a high probability of precipitation and detection of these latent defects when subjected to combined stresses in a HASS process.

The next step in bringing HALT and HASS methods to the manufacturer was to have all the key executives from engineering, manufacturing, marketing and finance together for a brief overview of the HALT and HASS methods. It was critical for all of the upper management to have everyone understanding that HALT and HASS would be a significant change from their current and classical reliability development methods. Not only would HALT and HASS require new thinking but also significant changes to manufacturing flow and new capital equipment expenditures. Having all the leaders of the company at a HALT and HASS overview presentation they heard the common questions and answers together, providing a simultaneous education, and preventing much of the misunderstanding and fear of change that the paradigm shift called HALT and HASS represent.

Since the manufacturer had good root cause records of failures, this data was presented to the executives showing why vibration and thermal cycling would be much more effective than their current process of burn-in, and would provide much higher probabilities of detecting it. The detailed history of the root causes of field failures also made it possible to provide a business case for HALT to HASS development.

With the history of failures, we were able to determine that 85% of the causes of field failures would have most likely been detected in a HASS process and the full payback for the purchase and set-up of a HALT/HASS chamber would be in 6 to 9 months.

The first product line to be subjected to HALT and HASS was a power supply. The particular model that was the first HALT to HASS candidate had two versions that had been shipping for approximately two years and so there were good records of the underlying root causes of field failures. We will refer to model A and model B in this case history.

During the previous two years of production, both models of manufactured units were subjected to a 72 hour burn-in that consisted of operating the units with a load on them in a closed test room. After four days of burn-in, the units were inspected and then shipped. The manufacturing process with steady-state burn-in screening had an annual warranty return rate of 5% during the last two years it had been shipping.

A.5.3 HALT Evaluation

The initial sample size for beginning HALT was three units. These units were easily accessed as they had been in production for some years.

The thermal protection circuits were manually disabled allowing the units to reach an inherent or raw stress operational limit and not the designed-in protection temperature set point.

Thermocouples were placed on key components for reference and comparison of temperature distributions between samples.

Each UUT had each step of HALT, first cold, then hot and then vibration. Each sample was subjected to HALT individually. Each UUT was inserted in the chamber and connected to electrical power and loads external to the HALT chamber to operate the unit at maximum rated power.

The HALT procedure began with cold step stress beginning at 10°C and decreasing in steps of 10°C with a dwell time of 10 minutes to reach thermal equilibrium before the next step. This process was repeated for hot steps, with each step being an increase of 10°C with 10 minute dwells. At each step the unit was power cycled and then run at full output load. Each sample was disassembled and inspected for any visually detected damage.

A.5.4 HALT Results and Product Margin Improvement

The first samples of the two models to be subjected to HALT were the Model A versions. The HALT of all the Model A samples resulted in thermal operation and destruct limits above 90°C and below −50°C, and vibration limits above 50 Grms measured at the vibration table. It was determined that there were no needed thermal or vibration margin capability to develop a good HASS process for the first of the two models.

The lower temperature thermal HALT operational limit was discovered to be consistent between samples in both model A and B, and the units would recover when the unit was warmed to ambient temperature. No low temperature destruct limit was found.

HALT evaluation of the Model B was a significantly different story. High temperature HALT of Model B samples resulted in the operational limit being the destruct or catastrophic failure limit. The three second model samples were found to have operation/destruct limits at 60°C, 75°C and 90°C. In all cases, the Model B unit's upper thermal operation/destruct limit was due to a power component that had a manufacturer's rating that was rated approximately 30% less power than the same component in the same circuit application used in Model A. Fortunately, the lower power component used in Model B and the higher power rated component used in the model A units were in identical packages, with the same mounting points and electrical contacts. This fact made it very easy to replace the lower rated power component in model B with the higher rated component used in model A. The cost difference between the higher rated component and the lower rated component was very small relative to the product costs.

The upper thermal destruct limit of the lower power rated component had a wide deviation of 30°C. The difference of 30°C is significant. With the small sample size, the mean upper operation limit (UOL) temperature cannot be determined. Yet, with this small sample size we can see that there is a wide distribution of the UOL in these samples. If this wide deviation in UOL extends to the larger production

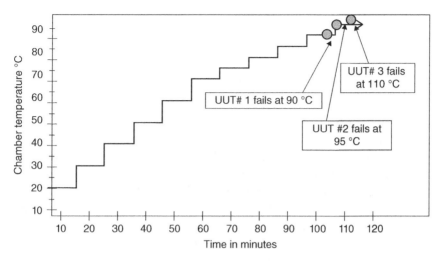

Fig. A5.1 New upper operating limits for model B units with high current component

quantities, some number of units may have operational problems at or near the operational end-use specification of 35°C.

When thermal HALT was applied to the model B units with the high rated power component replacement for the lower power rated component the UOL shifted significantly higher. The new UOL for the high current component is shown in Figure A5.1. The UOLs in the high current units also have a smaller deviation from the mean, which would lead to a lower distribution being less likely to extend into the worst case end-use thermal stress and potential failure.

A.5.5 HASS Development from HALT Results

The higher the thermal span and vibration levels, the more effective the HASS process is at precipitating latent defects, which of course is the central goal of HASS. A general rule of thumb and goal for the thermal span of a good HASS process has been minimum 100°C, although more is better. Another general rule of thumb for developing a HASS process, is to use 80% of the 90°C to −50°C range results in 110°C delta between high and low screening temperatures. The higher rated component in model B allowed for a >100°C HASS temperature cycling possible. The cost difference between the lower power rated and high current component was a few dollars, only a slight increase in the total cost of build.

A.5.6 Not All Management Agrees with Increasing thermal limits

It may be helpful to mention a common comment regarding the field relevance of failing the component in HALT. The engineering manager responsible for model A and model B argued that he had reviewed the failure analysis records of the field returns of the model B units and there were no incidences of the returned model B unit with the lower power rated component damaged as we had seen in the HALT evaluation. It was agreed that the component probably was not a cause of field failures, but without the higher rated component substitution in model B, the thermal portion of the HASS would be very limited and not as efficient. The manager agreed to accept the replacement of the higher rated component to allow 100°C delta for thermal cycles in HASS. The engineering change was made for model B units with the high current component substituted for the lower power rated component in the original design.

These limits allowed a thermal HASS to be run from 70°C to −40°C in both the models A and B. Both models were subjected to vibration HALT and were found to have an operation and destruct limit level at greater than 50 Grms, the maximum table input vibration. HASS vibration level is typically set at one half the destruct limit found in HALT. Although the units survived the 50 Grms input level, by default the maximum HASS level was set at 25 Grms maximum table vibration input.

The proposed HASS process was to begin with a stress regime of two thermal cycles from −40°C to 70°C while applying a modulated vibration up 25 Grms. Modulated vibration, discussed in Chapter 7, was applied to shift or sweep the harmonic peaks of vibration so that all potential defect sites resonate at their natural frequencies.

A.5.7 Applying HASS

The HASS process was implemented within the month after HALT had been completed and the engineering changes had been implemented for the manufacturing of model B.

To run the HASS process, each unit was placed in the HALT/HASS chamber and attached to input power with the output attached to a load. The loading and setup time was approximately 20 minutes and the run time for the HASS was 40 minutes.

The safety of screen (SOS) for the proposed HASS was applied. The SOS consisted of 10 applications on two new systems, the production process full load in the HASS chamber. After the SOS the units were completely functionally tested and visually inspected throughout the assembly. No flaws or damage was found and the units passed the complete final functional tests.

A.5.8 Resulting Warranty Rate Reduction

After completing the SOS on the model A and model B, the HASS process was applied to 100% of the production units.

The annual warranty return rate was 5% before the HALT/HASS process was applied. Within 3 months, the combined warranty return rate dropped by 90% to an annual failure rate of 0.5% on the two models of power conversion systems that were subjected to HALT and then the HASS process. They did an ROI analysis of the HASS process and found that they had an almost 3:1 return on the costs of the HASS process versus the pre-HASS warranty costs. This analysis was based on the conservative estimate that 75% of the HASS-precipitated failures would actually have resulted in field failures.

It has now been over two decades since this case history occurred and since that time the company now has HALT and HASS as a standard part of their reliability development process.

Index

Next Generation HALT and HASS: Robust Design of Electronics and Systems, First Edition.
Kirk A. Gray and John J. Paschkewitz.
© 2016 John Wiley & Sons, Ltd. Published 2016 by John Wiley & Sons, Ltd.

Wiley Series in Quality and Reliability Engineering

Reliability and Risk Models: Setting Reliability Requirements, 2nd Edition
by Michael Todinov
September 2015

Applied Reliability Engineering and Risk Analysis: Probabilistic Models and Statistical Inference
by Ilia B. Frenkel, Alex Karagrigoriou, Anatoly Lisnianski, Andre V. Kleyner
September 2013

Design for Reliability
by Dev G. Raheja (Editor), Louis J. Gullo (Editor)
July 2012

Effective FMEAs: Achieving Safe, Reliable, and Economical Products and Processes using Failure Mode and Effects Analysis
by Carl Carlson
April 2012

Failure Analysis: A Practical Guide for Manufacturers of Electronic Components and Systems
by Marius Bazu, Titu Bajenescu
April 2011

Reliability Technology: Principles and Practice of Failure Prevention in Electronic Systems
by Norman Pascoe
April 2011

Improving Product Reliability: Strategies and Implementation
by Mark A. Levin, Ted T. Kalal
March 2003

Test Engineering: A Concise Guide to Cost-effective Design, Development and Manufacture
by Patrick O'Connor
April 2001

Integrated Circuit Failure Analysis: A Guide to Preparation Techniques
by Friedrich Beck
January 1998

Measurement and Calibration Requirements for Quality Assurance to ISO 9000
by Alan S. Morris
October 1997

Electronic Component Reliability: Fundamentals, Modelling, Evaluation, and Assurance
by Finn Jensen
November 1995

Printed and bound by CPI Group (UK) Ltd, Croydon, CR0 4YY

27/10/2024

14580207-0003